JUST KIDDING!

A to Z Designs for Kids & Kidults

Published & edited by Viction:ary

JUST KIDDING!

A to Z Designs for Kids & Kidults

First published and distributed by
viction:workshop ltd.

viction:ary™

viction:workshop ltd.
Unit C, 7/F, Seabright Plaza, 9-23 Shell Street,
North Point, Hong Kong
Url: www.victionary.com   Email: we@victionary.com
www.facebook.com/victionworkshop
www.twitter.com/victionary_
www.weibo.com/victionary

Edited and produced by viction:ary
Concepts & art direction by Victor Cheung
Book design by viction:workshop ltd.
Cover images: Nokia Cover by Jean Julien

©2013 viction:workshop ltd.
Copyright on text and design work is held by respective designers and contributors.

All rights reserved. No part of this publication may be reproduced, stored in retrieval systems or transmitted in any form or by any means, electronic, mechanical, photocopying, recording or any information storage, without written permissions from respective copyright owner(s).

All artwork and textual information in this book are based on the materials offered by designers whose work has been included. While every effort has been made to ensure their accuracy, viction:workshop does not accept any responsibility, under any circumstances, for any errors or omissions.

ISBN 978-988-19439-6-5
Printed and bound in China

ICON INDEX

JUNIOR GAME

Safety might be a priority but equally important is how well a design can stimulate a child to think and develop skills essential to their growth. These projects will just meet their needs and want. They are all grounded in all the care and love we can give our kids.

KIDULT'S LOVE

Happiness can be as simple as that. Taking on a fanciful notion or a bold palette, these projects are there to put that primitive urge to dream and learn back into adults' life, like we used to feel in our salad days.

FUN FOR ALL

Children and adults look at things in dissimilar ways, but this is where their common ground lies. Be it a character or a chair, their imaginative quality piques little one's curiosity and tickles grown-up's fancy. These projects bring a simple pleasure to all age groups.

CONTENTS

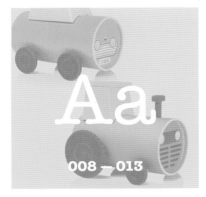
Aa
008 — 013

Bb
014 — 029

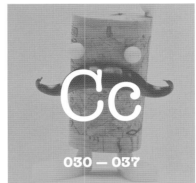
Cc
030 — 037

Dd
038 — 049

Ee
050 — 053

Ff
054 — 061

Gg
062 — 065

Hh
066 — 075

Ii
076 — 085

University Libraries
Carnegie Mellon University
Pittsburgh, PA 15213-3890

Jj

086 — 087

Kk

088 — 095

Ll

096 — 097

Mm

098 — 123

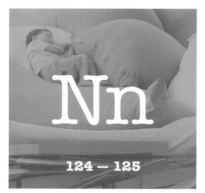

Nn

124 — 125

Oo

126 — 135

Pp

136 — 161

Qq

162 — 163

Rr

164 — 171

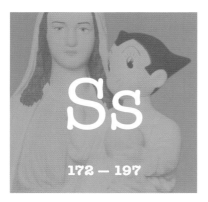

Ss
172 — 197

Tt
198 — 207

Uu
208 — 209

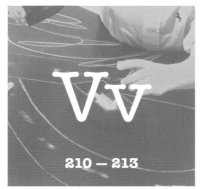

Vv
210 — 213

Ww
214 — 221

Xx
222 — 225

Yy
226 — 227

Zz
228 — 231

Good design is about filing a need in a sensible way. Aside from telling a good story and combining good colours, good design has to be comfortable.

A good playground should inspire kids to move. The swing and the ropeway are classical playground elements where you can feel the joy and the tickle in your stomach just by looking at it. A GREAT playground is a place that kids are not able to figure out just by looking at it. They have to explore it. When they are running or climbing through the playground there is not only one right way. They have to consider a lot of options and paths assessing their mobile skills and safety. This creates continuous movement.

Kids want to be challenged and they want to belong. But on the playground there are a lot of different groups with different needs and skills. Girls and boys, young and old, disabled and normal skilled kids are not the same and if you try to create a playground where everybody can go everywhere you are sure that everybody will be bored. But I have seen girls climbing, boys kissing and blind kids playing football. Good playgrounds are usually focused on a specific target group, but these are not necessarily defined by age or gender but by skills, needs and interests. Children measure their progress in life by growing up and leaving things behind and this is what inspires them to explore.

In MONSTRUM we pay special attention to making our playgrounds reasonably safe. We want to combine adventure and danger with the possible risk and protection from being injured. It is our foremost goal to keep focus on safety from initial concept to finished product. With this in mind we try to make playgrounds that look dangerous but are safe. Our goal is to get the child to take responsibility for his/her own safety.

The world is a truly fantastic, colourful and dangerous place for kids to grow up in. The playground has to be an equally inspiring alternative where kids can learn to assess risks in a safe environment.

Children are not interested in design. Good or bad. But like most people, children feel the impact of it. Unlike art you don't have to understand design to enjoy it.

This book gives me design that makes me feel happy but not stupid because I don't understand it. I want to be educated, happy and understood.

FOREWORD

by
Ole Barslund
Nielsen,
MONSTRUM

Ff

FOREWORD

by
Tim Durfee &
Iris Anna Regn

The notion of childhood has changed radically over the last 50 years. Whereas childhood in industrialised nations was previously considered a miniature adulthood, in just a few generations, it became an established and recognised period within a person's life, and children holders of terrific potential for their and our future.

Accordingly, one of the beliefs that we as a society have adopted into the mainstream from 20th century psychologists, Jean Piaget and Lev Vygotsky amongst others, is the essential importance of 'play'. Play is considered so critical to human development that it was recognised by the United Nations High Commission for Human Rights as a fundamental right of every child. Subsequently, the National Institute for Play in California states as their mission to "[unlock] the human potential through play in all stages of life [and use] science to discover all that play has to teach us about transforming our world." Once the sole domain of children, play has more recently become significant and motivational for us at any age.

All of which brings us to the innovative and playful collection of artefacts in this book. *Just Kidding!* underscores the contemporary relevance of taking a kid-like stance and of incorporating play — of which humour is considered a grand subspecies — into all human interaction. As both a designer and a parent, what fascinates and encourages me about this work is that it documents and reveals a unified, specific, quirky optimism that is running through our current period. We now believe not only that children can learn from adults but we designers also honour and concretise the assumption that we can and want to learn from children.

Just Kidding! is a compendium that is diverse enough to take the idea of kidding from a specific moment in design to a more universal shift in the way we think; hence the artefacts in the book — symbols of our manifold human interactions — range from clothing and furniture, photography and interior design, to food packaging and actual toys. The book also features contemporary design studios of vastly different sizes and global locations, most of whom, interestingly, are multi-disciplinary but do not specialise in design for children. The extreme breadth of the work is surprising in another aspect as well — some of the work is commercially available, some appears in private institutions, some can be commissioned bespoke, some is one-off. What all the projects do, however, have in common is that they utilise strategies of play to achieve layered meaning and expanded engagement.

In *Just Kidding!*, to be child-like, kid-like, has been treated with obvious respect, one that has the potential to take us from the ridiculous to the sublime.

Aa

"ANIMALS"

Air Heads is a horde of party animals awaiting users to complete their faces. Available in six colours, each balloon set reveals a creature, ranging from bird, bear, koala, tiger to rhino and pig that comes with a wand, matching paper features and stickers for eyes and patterns. Features like nose and horns come in the pack as pre-cut paper attached with sticky pads. The paper flap system means no glue required throughout the make.

Project / Air Heads
Category / Toy
Design / Héctor Serrano
Client / npw

"APARTMENT"

Architect Felix Grauer has a new way to build but at a much smaller scale. Capitalising on Lego system's brick-building properties, especially Lego Technic blocks which contain holes, Grauer turns the colourful studded bricks into flexible and practical key fobs with split rings. Users can easily bind and unbind keys according to need. A matching Lego building plate will perfect the idea as a "hanger" where spare keys can be mounted and kept.

Project / Lego DIY Key Hanger
Category / Home accessories
Design / Felix Grauer

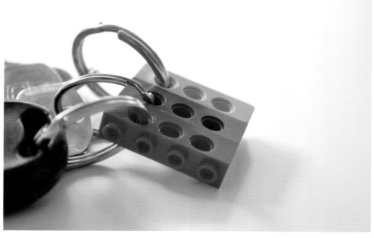

"AUTOMOBILE"

The Tube Toys series will probably remind you of the fun of creating simple toy with a toilet paper roll. With the four vehicle-themes — fire truck, tractor, train and car, each of the tube contains stickers and all components necessary to complete a push-along motor. The packaging tube doubles as the vehicle's body with die-cut slots to locate axles, wheels and cabs.

Project / Tube Toys
Category / Toy
Design / Oscar Diaz Studio
Client / npw

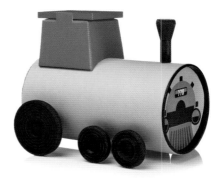

"ALPHABET"

Everybody receives a name after they are born. As often as not, names carry meanings inherited from cultures, stories, myths and notable bearers in history. Meaning of Names is a deck of beautifully illustrated cards where personal names are specified along with cartoon strips. Picturing each alphabet in regards to the meanings of relative names, the set presents a collation of interesting findings we could not have imagined.

Project / Meaning of Names
Category / Card sets
Design / Young & Innocent

A-Z

Bb

"BUBBLE"

Bubbles are visually inviting but delicate. Casting back the surrounding while they take shape, these iridescent spheres burst on its own or at a touch. Intended to be a lamp that constantly transforms with the most ephemeral of lampshades, Surface Tension Lamp produces its own cover by blowing a soap bubble around the LED light source. Once the bubble breaks, a new bubble starts to form.

Project / Surface Tension Lamp
Category / Lighting
Design / Front, Loligo
Client / Booo
Photo / Pär Olofsson

"BALLOON"

Balloon brings smiles as it floats and bounces, and takes them away as it deflates, but Balloon X Lamp is an aggregate of hope and warmth for as long as it takes. Taking the shape of balloons with wires as their string, this wall lamp of Haoshi's Childhood series, ties us to a time when life was filled with sweet memories, dreams and gentleness.

Project / Balloon X Lamp
Category / Lighting
Design / haoshi Design
Client / MANBO KOKUSAI, ICI

"BONBON"

Rainy days do not necessarily convey only blue and grey. In three "flavours"— Green Apple, Mint and Vanilla Strawberry, these cheerful candy-like crook handle covers are no mere eye candy. Not only do they add colours to the dull dark days, but also offer comfort to your grip and help distinguish your own umbrella right away in an umbrella stand. Candy in the Rain is a collaboration between Bob Foundation, Danke and Nol Corporation.

Project / Candy in the Rain
Category / Accessories
Design / Bob Foundation
Client / Danke

"BLOCKS"

Toy blocks are among the first tools we use to learn to analyse a world made up of visible and invisible shapes. PlayShapes is a unique modular set of 74 wooden units comprised of not only simple geometric shapes, but also a sprectrum of arches, quadrilaterals, sectors and eyes of various style and sizes which can be arranged into hundreds of flat or three dimensional compositions, like animals, faces, planes, cityscapes and many more. Colourfully tinted on one side, these shapes are beautifully crafted from environmentally friendly, durable rubberwood.

Project / PlayShapes

Category / Toy

Design / miller goodman

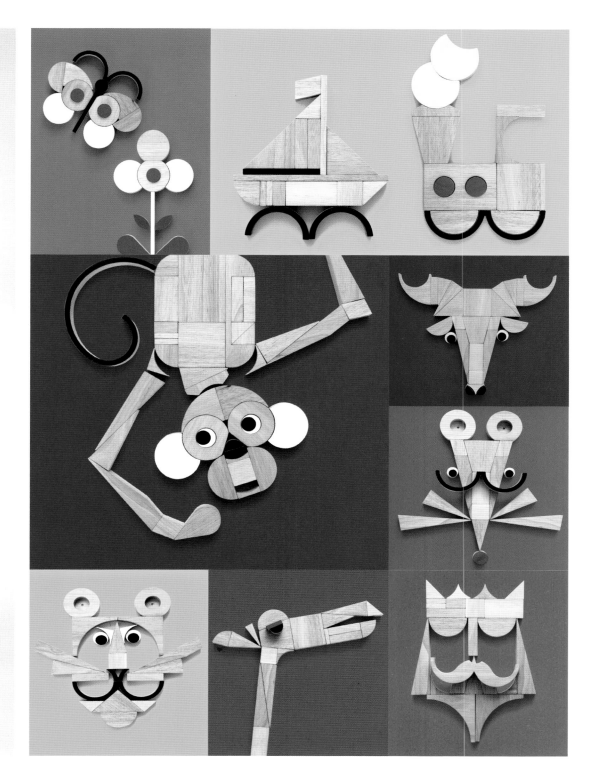

"BARBER"

Create hairstyles by simply erasing! Each bearing a different face surrounded by a square of colour, these erasers ask users to shape their heads by gradually rubbing them against your notes or desk's surface. Users can decide what makes them look best with preset fringes, eyewear and character as it shows on their faces. There's no going back but sometimes mistakes are good for a laugh.

Project / Rubber Barber
Category / Stationery
Design / Megawing

"BAT"

What could bats do in a dark, enclosed closet? Although reckoned as nocturnal creatures, Bat Hangers can always spread their wings to hold your clothes, while folding them at other times. But no matter what, you'll find them hanging upside down most of the day. Simple yet cleverly designed, the utterly black hangers feature teeth at the turn of the wings near the joints to prevent slipping as they move.

Project / Bat Hanger
Category / Home accessories
Design / Veronika Paluchová

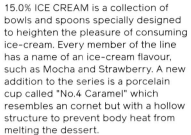

"BLISS"

15.0% ICE CREAM is a collection of bowls and spoons specially designed to heighten the pleasure of consuming ice-cream. Every member of the line has a name of an ice-cream flavour, such as Mocha and Strawberry. A new addition to the series is a porcelain cup called "No.4 Caramel" which resembles an cornet but with a hollow structure to prevent body heat from melting the dessert.

Project / 15.0% ICE CREAM
Category / Tableware
Design / TERADADESIGN
Client / 15.0%

"BISCUIT"

Dinosaur biscuits can now get up and walk on four legs — or perhaps not for a T-Rex. Oddly shaped, these plastic cookie cutters, in a set of three, help bakers to press out dinosaur body parts from the dough to create biscuit pieces required to complete a 3D dino model. The collection has cutouts for four dinosaur species — Triceratops, Stegosaurus, Brachiosaurus and T-Rex.

Project / 3D Dinosaur Cookie Cutters

Category / Kitchen utensils

Design / Suck UK

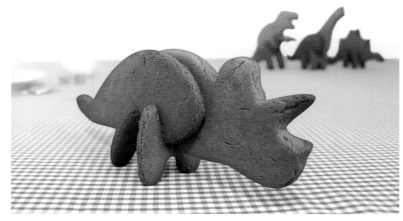

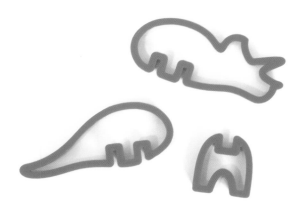

"BAKERY"

Sapore dei Mobili (Furniture Tasting) is not only about molding cakes into cute little lifelike furniture so that you can have fun building an edible living room, it also comments on the velocity of furniture production today. Bake as much as you can eat! With different recipes, the pan create a variety of chair, coffee table, cabinet, lamp and vase designs to go with sprinkles and jam.

Project / Sapore dei Mobili
Category / Kitchen utensils
Design / Rui Pereira, Ryosuke Fukusada
Photo / Alfredo Dante Vallesi
Special credits / Matias Baldassarri, ZooiLab

"BENCH"

Whoever owns a SEESAWseat at home deserves to be proud. Easily convertible into a teeter-totter or back into a bench in just a few steps, it caters to grown-ups or children whoever have a constant impulse to play or just take a seat to enjoy summer breeze or a chatter between friends. The transforming takes place quickly — it's a simple matter of sliding in and out its two parts and shifting three pins. Suitable for indoor and outdoor use.

Project / SEESAWseat
Category / Seating prototype
Design / DvanDirk
Photo / Ivo Knubben

"BEAR"

Teddy bear is always a cherished childhood keepsake. Where history and its varied interpretations over time have transmuted its endearing nature into our most treasured memories, Mateo the Bear borrows these emotional values and brings them to your quotidian activities as a delicate pencil holder.

Mateo is handmade by Mexican artisans with sheep leather and articulated arms and legs, so have fun with it. Help him strike a different post every day as long as your pens stay within.

Project / Mateo the Bear
Category / Stationery
Design / Christian Vivanco Design Studio

Cc

"CUP"

If coffee or tea is not enough a break for the day, how about a minute-trip to the amusement park? Playing on reflections, each tea set introduces a ride or attraction in a park which completes only if the plated cup goes with the matching saucer, which can spin without upsetting the drink. Release the ghosts, set off the carousel and let out the fishes. Power up the games with a push.

Project / A Little-round-round Land
Category / Tableware
Design / biaugust CREATION OFFICE

"CASE"

Sleep tight, for they guard your stories. Called GUARDACUENTOS, Spanish for "tales guardian", these shelves depict an array of merry forest animals who incarnate themselves in furniture to protect storybooks while kids go to bed. The family has eight members, including an owl, a rabbit, a fox, a beaver, a bear and a hedgehog, as part of a kid furniture collection, "The forest at home".

Project / GUARDACUENTOS
Category / Bookshelf
Design / Menut
Client / Mirplay
Photo / Carlos Moreno, Carlos Lluna

"CARPET"

Composed of a rug and a cushion seat, OLI offers soft texture and a pleasant touch like the layered fallen leaves in a park. To recreate that feeling, the rug is produced as a stack of felt leaves of various shades. OLI is part of the "The forest at home" kid furniture collection which also features bookshelves, a coat rack and a chair.

Project / OLI
Category / Home accessories
Design / Menut
Client / Mirplay
Photo / Carlos Moreno, Carlos Lluna

"CORK"

Corkers is a collection of plastic pins which you can push into cork stoppers to create little animal characters at the dining table. Each kit supplies limbs and features characteristic of one of the six animals the series depicts, but the parts can also intermingle to create peculiar genera. The animal range features monkey, deer, buffalo, bear, bunny and crow.

Project / Corkers
Category / Toy
Design / Reddish, Monkey Business
Manufacture / Monkey Business
Photo / Dan Lev

"CHALKBOARD"

Mavalà originates as a gift to the designer's son. Handmade from recycled wood with broad magnetic surfaces and a coat of water-based paint, the furniture set fulfils the functions of an enduring learning instrument for young children and a handy message board for grown-ups. The table can remain in use as a stool with larger desks as children grow bigger.

Project / Mavalà
Category / Furniture
Design / Anna Licata

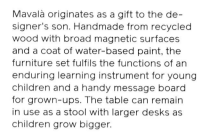

Dd

"DOODLE"

Home furniture often defines the person who lives in the space. Where children's furniture are usually finished designs, Inside Out Lamp reserves space for children to put the finishing touches to its look. Using double layer polycarbonate sheets, the lampshade allows users to draw on both sides and wipe off old sketches to start anew. LED light source ensures painters are safe to draw inside while the light is on.

Project / Inside Out Lamp
Category / Lighting
Design / Lin Yu-Nung

"DREAM"

Bed is where dream and reality switch places. Underneath these sheets you will dream far beyond the stars. With high resolution photographic prints transferred onto pristine white percal cotton, SNURK's themed duvet cover sets will send dreamers to their ideal destinations in no time. Notes for attention: Astronaut is a faithful depiction of the official space suit design archived by Dutch Space Expo Museum, and Trampoline does not change the fact that a bed is not made for jumping.

Project / Astronaut, Princess, Trampoline
Category / Beddings
Design / SNURK, Theo van der Laan
Photo & Styling / Tim Stet, Kim de Groot

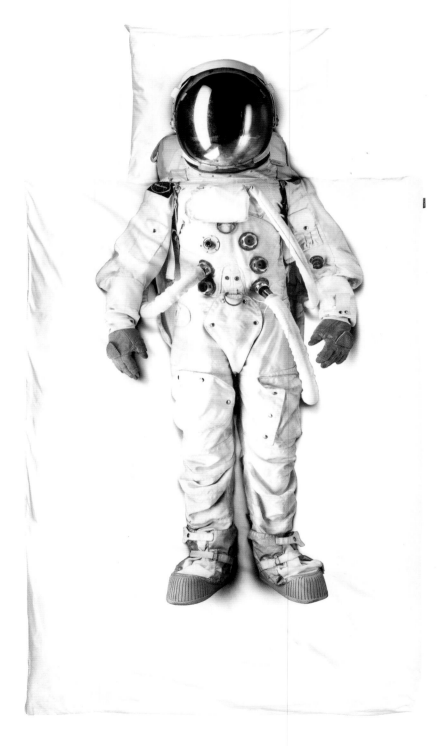

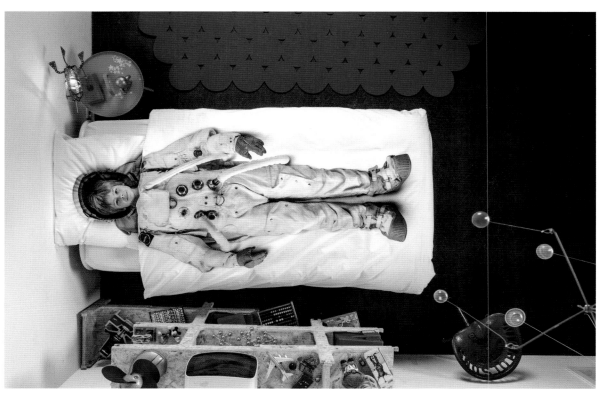
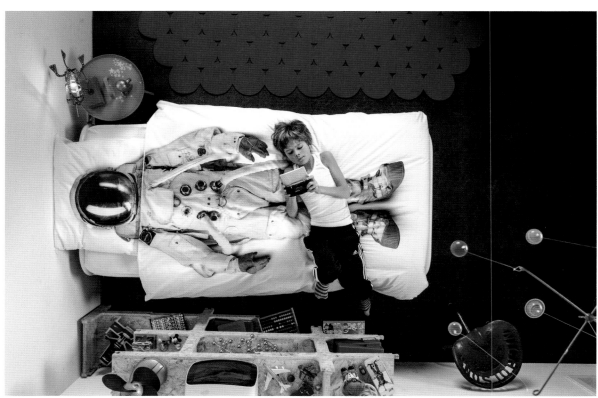

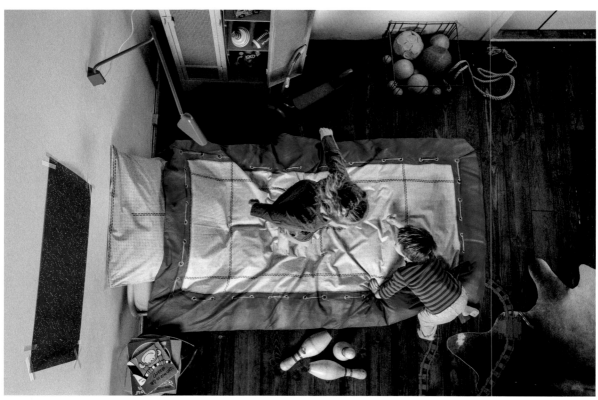

"DRAWING"

Getting one's work mounted for public viewing is an artist's great source of pride. Like Post-it notes, with adhesive on the back, Sticky Sketch saves the troubles of framing, trimming and fixing — a problem that all kids face when they wish to pin up their work. Sticky Sketch comes as a sketchbook and it is easy to attach to and remove from the wall.

Project / Sticky Sketch	
Category / Stationery	
Design / Yeongkeun Jeong	

"DESK"

Anyone standing in front of Guillaumit's Animal Drawing Table will immediately be absorbed into the world of imagination. Transferring characters of a bird, fish, pony and serpent into bold, geometric forms in rigid colours, these calm-looking creatures bring the French illustrator's characters to life as a sculpture and functional furniture, inviting kids and adults to forget themselves in illustration. Various body parts of these cute friends have been devised to stock drawing paper, pencil sharpeners and pencils as hair or spines.

Project / Animal Drawing Table
Category / Furniture
Design / Guillaumit
Manufacture / Espace Espinoa, La Gaîté Lyrique
Special credits / Pictoplasma

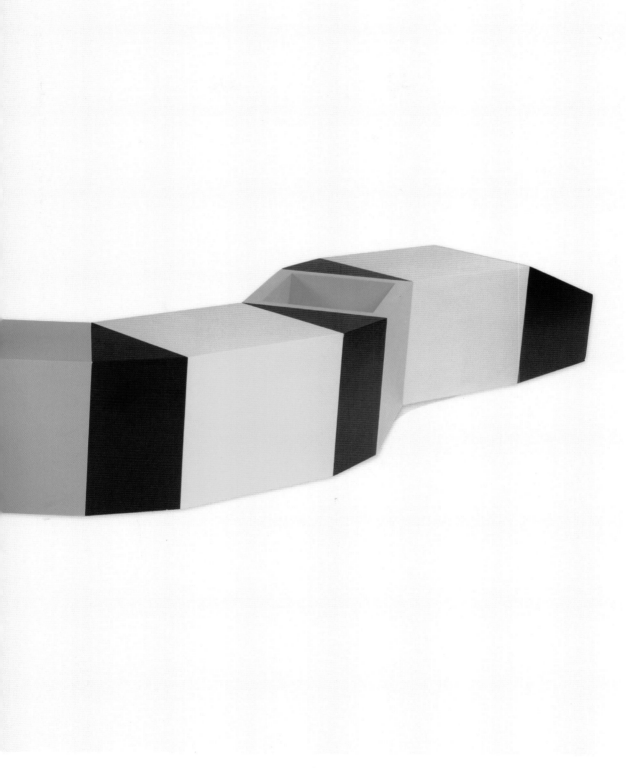

Ee

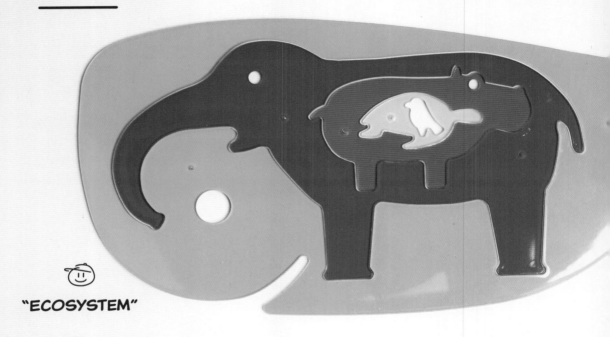

"ECOSYSTEM"

Although these animals do not really feed on each other, Hungry Animals tell a story about how animals interconnect within the ecosystem. All integrable into the body of a whale, the colourful plates can be separated as stencils with which children can use to draw and colour. By putting them back together, the puzzle reveals the relative proportions of five animals, such as sea turtles and hippos, which are vulnerable animals that require attention.

Project / Hungry Animals
Category / Toy
Design / Oscar Diaz Studio
Client / Natural Products Worldwide

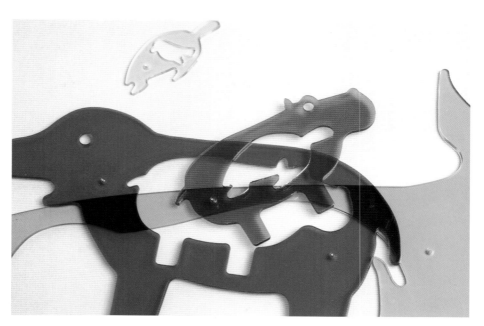

"ELEPHANT"

CALEPHANT is a 'planter' which grows joyous leaves on an elephant's back. Holding 12 calendar pages, each attached with a small plant of a different scale and colours, users will witness its growth and colour changes as they update the sheets every month. CALEPHANT can stay on the desk as a cheerful pot even after the calendar expires at the end of 2012.

Project / CALEPHANT
Category / Calendar
Design / RMM

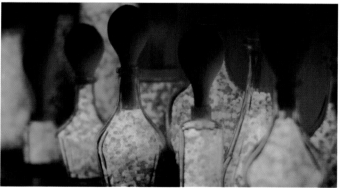

"EMERALD"

Courtoise is your night guide. With marbles made out of glass and phosphorus, held in vintage stoppered glass containers, Courtoise lights up the hallway and scares off monsters as it glows and diffuses a poetic light in the dark. While the glow fades away during the day, the glass beads blend with the interior in the colour of roseate, green and white.

Project / Courtoise
Category / Home accessories
Design / Maisonnée
Photo / David Lemonnier

Ff

1

"FAIRY TALE"

Every one of us grew up reading fairy tales. The fantasy and moral teachings remain as a happy influence in our different walks of life. Originally created for Tokyo Designers Week 2011, Once Upon a Time depicts an array of folk stories in modernised forms. The project includes The Kingdom as lampshades that resemble the icon of castles[1]; a pencil sharpener called "Honest Pinocchio" whose nose never really extends[2]; and a matryoshka doll that condenses the story of Little Red Riding Hood into a set of nested dolls[3].

Project / Once Upon a Time
Category / Lighting, stationery, toy
Design / Pistacchi Design
Client / (Honest Pinocchio) Kikkerland Design Inc.

"FINGER"

Let the fingers do the talking. With the variety of paper puppets and stickers the London design house produced for npw, children can create their own play with farm animals and fairy-tale characters jamming into the stage anywhere they go. Finger Puppets are pre-cut from card and easy to assemble, enabling mouth movements as users bend their fingers. Finger Tattoos are adhesive prints, available in four themes like monsters and safari.

Project / Finger Puppets & Finger Tattoos
Category / Toy
Design / Héctor Serrano
Client / npw

"FLAG"

Step to the World is a memory game in the shape of sock. Like socks which goes in pairs, the game asks children to pick out sets of two, of which the right one bears a country's national flag, name and a greeting in the language their people speak, and the left shows flag elements accustomed to the sock's form. Each set contains 15 pairs of socks made from beech veneer, with plain ones inclusive for your own design.

Project / Step to the World
Category / Games
Design / Kukkia

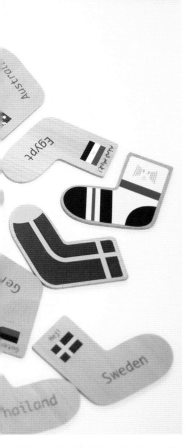

"FOOTBALL"

What if it rains? What if you have an unstoppable passion for football but would rather watch sports on the sofa than actively play it on the field, like Emanuele Magini, the designer of Lazy Football? Although at a much smaller scale of the game in terms of players involved and the size of the field, the comfortable home version with the goals incorporated into chairs demand agility and physical dexterity to score. The game can be incredibly intense if players shorten the distance between the goals.

Project / Lazy Football
Category / Furniture, games
Design / Magini Design Studio
Client / Campeggi
Photo / Ezio Prandini

Gg

"GEOMETRY"

With the inclusion of wood pegs, Dowel Blocks is about to free imagination of building by removing the restrictions of gravity and coordinating shapes. Featuring holes on five sides, these polyester-coated plywood units can be connected three-dimensionally to create unexpected structures which are then safe to carry around. The multi-colour finish adds a sense of glee to its final look.

Project / Dowel Blocks
Category / Toy
Design / Torafu Architects
Manufacture / Ichiro
Photo / Akihiro Ito

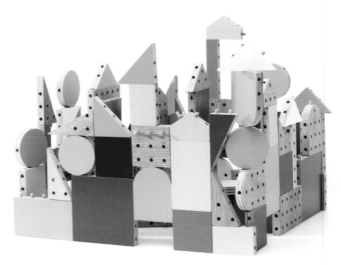

"GLOBAL-WARMING"

The rocking 'bears' is a sad-yet-beautiful love story between a polar bear and a grizzly bear. Depicting two baby bears walking falteringly on the melting ice, the story originates as the globe continually warms up, melting the Arctic homes of polar bears while forcing the North American brown bears to move up for better living conditions. 'Bears on Melting Ice' was originally a gift for the designer's daughter.

Project / Pizzly Bear
Category / Toy
Design / Masahiro Minami Design
Client / Huzi Design

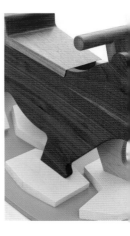

Hh

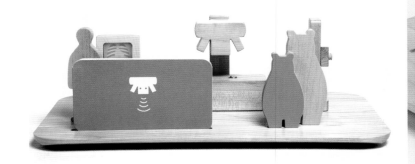

"HOSPITAL"

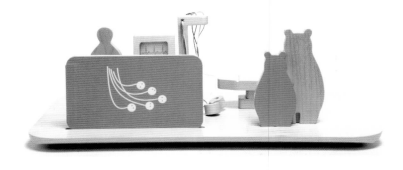

Hospitals might not be a very popular destination for children. Often overwhelmed by sickness and anxiety brought by unfamiliar environment and procedures, little patients can almost always only sense and recall fear as they look back on their last trip to clinics while they need to prepare for the next one.

Novel Hospital Toys feature illustration books and toy models of machines for CT scan, X-ray, echo, ECG with which parents could use to informing their babies of what to expect during their stay. Every machine can give light or sounds to resemble how these machines operate.

Project / Novel Hospital Toys
Category / Toy
Design / Hikaru Imamura

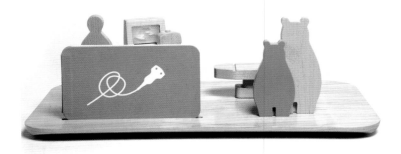

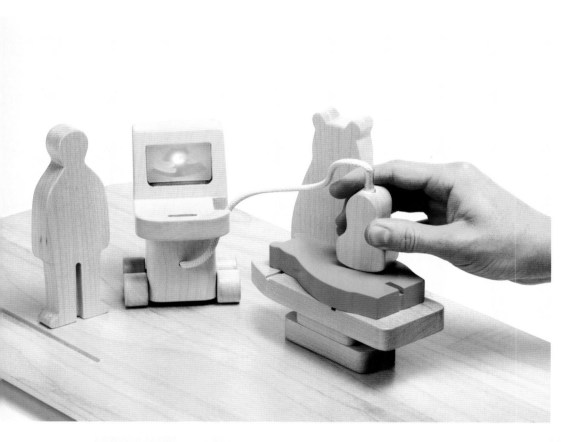

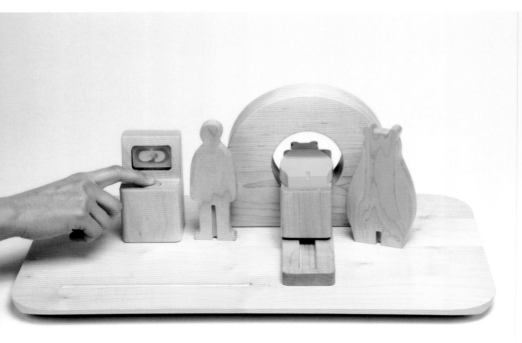

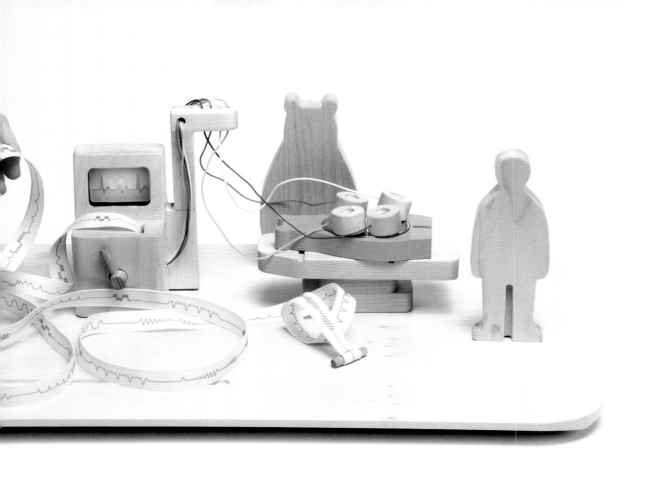
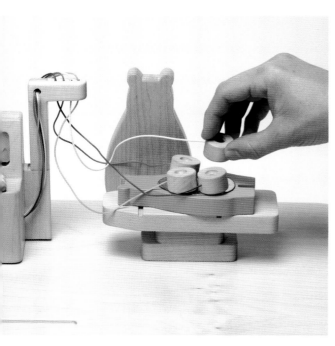

"HOUSE"

Made of recycled cardboard, Casa Cabana Collection is a toy house and furniture collection comprised of houses, rocket, trojan horse and small furniture pieces. Kidsonroof as well encourages children to be creative about these environmentally-conscious structures which come with printed silhouettes of animals and scenes. By drawing and painting, children can make the little house and rocket become their very own territory.

Project / Casa Cabana Collection
Category / Toy
Design / Kidsonroof

"HOOF"

Ready to disguise yourself by leaving animal footprints behind wherever you go? Perfect for a walk in the beach or a sand pit, these toy sandals is a modification of traditional geta, with the base boards replaced by shapes that can reproduce hoofprints on soft grounds. The range features the feet of five animals, including cats, geckos, monkeys and owl.

Project / Ashiato Sandals
Category / Toy
Design / Kukkia

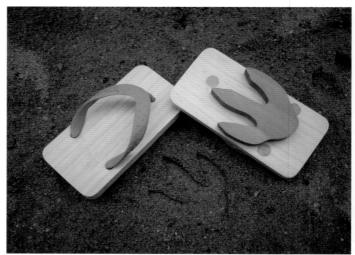

Ii

"ICE POP"

Referencing a classic product pack shot, the Popsicles series creates an imitation of a common object through a purposeless addition. Both products, as a sponge or an ice-lolly, represent different aspects of everyday life and are recognisable in their own right. They are morphed into a fictional replica creating a visual double take and a dysfunctional bi-product.

Project / Popsicles
Category / Photography
Design / PUTPUT

"IMP"

These are no simple masks! Densely covered in kilos of marshmallows, gummies and rock candies of all sorts of colours, these magical masks will take their wearers to a fantastical reality, like the ones the Shamans own. The series was created as a visual exploration with Dutch candy company Jamin and its sweets, for Graphic Detour, curated by Erik Kessels. They have made their first public appearance at the Museum of the Image in Breda, the Netherlands, in 2011.

Project / Masks & Sweets
Category / Artwork
Design / Damien Poulain
Client / Graphic Detour, Museum of the Image, Breda

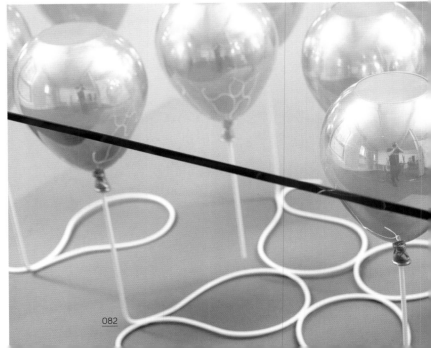

"ILLUSION"

Is it the balloons which are propping up the solid glass panel or the glass table holding up the helium-filled balloons? Fashioned out of toughened glass, metal resin composite for the balloons and toughened steel rods of various length, UP coffee table conveys a playful trompe l'oeil that defies gravity and unites function and art. The table is produced with an edition of 20, all by hand in the UK.

Project / UP
Category / Furniture
Design / Duffy London

Jj

"JOURNAL"

If you wonder why these notebooks have to wear bands, take it off and you will reveal funny faces that show an urge for learning or satisfaction for being your company in school or at work. Called Incognito, these notebooks are held by a band that conceals a character's face. Die-cut grips are made to ensure their real identities are not revealed at an inappropriate time.

Project / Incognito
Category / Stationery
Design / Ozan Akkoyun
Creative direction / Barış Akbaş
Client / Happily Ever Paper

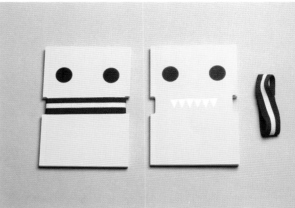

Kk

"KEEPSAKE"

Call it a toy capsule or a time capsule. Initially designed for a kidswear brand, drawing ideas from capsule toys well-liked by children and adolescents, Capsule Lamp is transformable, whether it carries zero capsule or with its full gear on, be left vacant or filled. Whatever you load into the containers, the objects together crystallise your life and state of mind at a particular time.

Project / Capsule Lamp
Category / Lighting
Design / Design Systems Ltd
Client / ACTIF

"KINGDOM"

Now playing house can be more true to life. With Designer Mama, "house" literally comprises an interior and a hodgepodge of designer furnishings, with Eames storage unit, George Nelson clock, Thorsten van Elten Antler coat rack, Mia Hamburg's Shuffle Table and surely a Polish folk cupboard and "Fawn" children stool designed by Olgierd Szlekys and Wladyslaw Wincze in the 1950s.

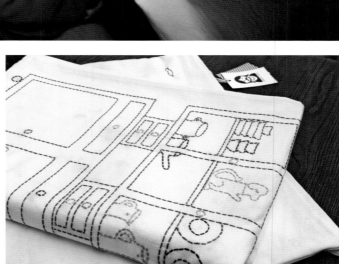

Project / Designer Mama
Category / Home accessories, toy
Design / Pani Jurek, Gang Design
Photo / Barbara Kaja Kaniewska

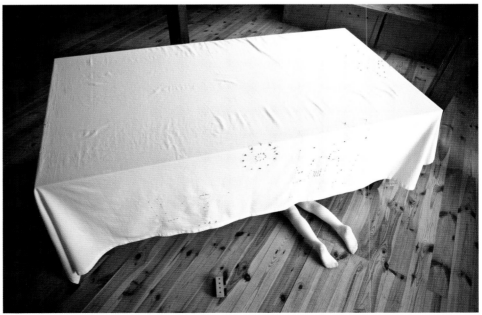

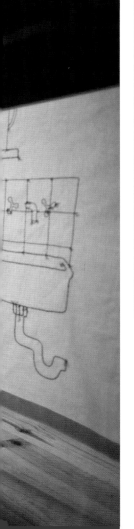

"KIT"

Atelier Book Chair is a portable studio with all that a little artist would need to paint boxed into one case. Made from Japanese cypress, a durable material sourced from Nishiawakura village in Okayama prefecture, Japan, the kit holds pouches for painting equipments and a trapezoid-shaped plank as the seat which can be slotted onto the case as it folds out as the base. Accessories such as the grip, name tag, stripes and the latch are all fashioned out of leather.

Project / Atelier Book Chair
Category / Stationery
Design / Kana Nakanishi
Art direction / Chikako Oguma
Client / Oiseau., Inc
Manufacture / Masayuki Oshima
Photo / Asaco Suzuki

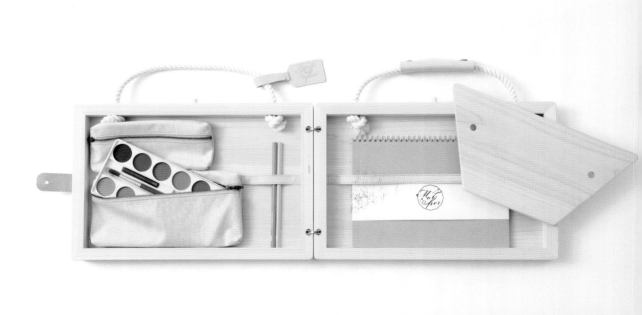

Ll

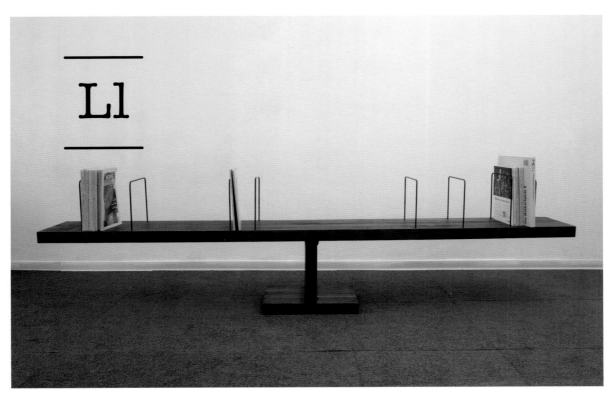

"LIBRARY"

Is Kafka truly heavier than Vogue's latest issue? Do you read as much classics as Japanese mystery manga? While every book tells its own story, every book has its own weight. By playing with balance, See-Saw bookshelf as part of BCXSY's PLAY! collection visualises the breadth of our home libraries as we set our books on either end.

Project / See-Saw Bookshelf
Category / Bookshelf
Design / BCXSY
Photo / Herman Mertens

Mm

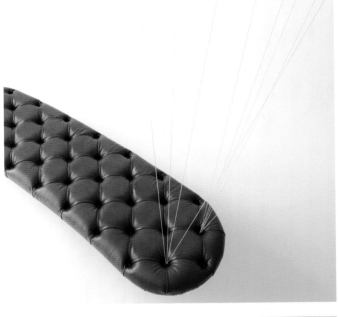

"MAGIC"

Can one really travel by cluster balloons? Visually inspired by the feeling of floating that a young boy experienced in the movie, Le Ballon Rouge (1953), Balloon Bench is made to hover steadily in the air with bunches of balloons tied to its ends.

Certainly filled with helium but not physically lifting the seat, the balloons conceal the anchor points in the ceiling to create an illusion which many want to be real.

Project / Balloon Bench
Category / Furniture
Design / h220430
Photo / Ikunori Yamamoto

"MUFFIN"

The texture of these colourful range of oversize muffins is no difference to the ones freshly delivered by a radical bakeshop, only bigger and tougher so you could sit or rest your feet on them. These leather-cushioned stools are available with a choice of "flavours" and "toppings (buttons)", with compartment for toys and memories inside the solid wood cups.

Project / Mini Muffin Pouffe
Category / Furniture
Design / Matteo Bianchi Studio
Manufacture / Linden Bauer
Photo / Rei Moon

"MOUSTACHE"

Chocolates have a special power that brightens people up, whether it comes as a drink, cake slices, little cubes, puddings, in a dessert or a savoury dish, in a loaf or as a moustache on a stick. In a variety of shapes and flavours — from the more traditional dark chocolate and white chocolate, to the more unorthodox caramel- and strawberry-fused style, Mr. Chocolate is an ideal disguise for clean-shaven chocolate lovers of all ages.

Project / Mr. Chocolate
Category / Confectionery
Design / Diego Ramos Studio
Client / Chocolat Factory
Special credits / Concha Carrascosa Lezcano, Otto Capo Barjau, Raquel Quevedo, Kentaro Terajima

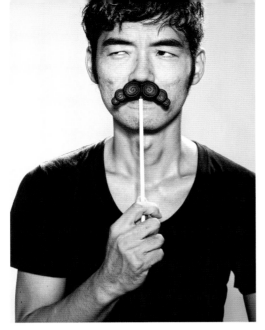

"MOBILE"

Originally designed to connect people, mobile phones today rather occupy users and set them apart with games, videos, music and speedy news update. How about restoring its primitive function with a basic key pad? Each in a distinctive colour and body shape, bearing funny faces, the not-so-conventional phone covers appeared at the Nokia world exhibition 2011.

Project / Nokia Covers
Category / Accessories
Design / Jean Jullien
Client / Nokia
Manufacture / Toufan Hosseiny
Photo / Pelle Crepin

"MASK"

Peekaboo! Called Bu! Indian with two openings at the centre over an Indian illustration by Elisabeth Dunker aka Fine Little Day, the round cotton blanket lets babies peek before giving out a loud "Boo!" while keeping them warm on a trip. The Bu! blankie line has recently added three new motifs, featuring teddy, pirate and lion, respectively drawn by Lisa Grue, Will Broome and Elisabeth Dunker. All prints are available in two colours matched by different colour edges.

Project / Bu! Indian
Category / Toy
Design / Little Red Stuga
Illustration & Photo / Elisabeth Dunker

"MAMMALS"

Wonderfully black, Animal Chairs is a herd of tame and gentle animals that highlight the pleasure of sitting as much as the unique quality of each material used. A harmonised play of form and craftsmanship, the seats showcase a balance between sophistication and imaginative quality, furniture and art. The series is an exemplar of artistic crossover involving skilled carpenters, blacksmith and tailors from Taiwan.

Project / Animal Chairs
Category / Furniture
Design / biaugust CREATION OFFICE

"MIRROR"

Auguste is the door to the world where you can join Alice and the peculiar creatures for a daring adventure. Crested with a pair of big porcelain pointed ears, Auguste also features a fleshy face that befits all types of face. Move over to Auguste and put up a smile, you will find yourself dressed up like a white rabbit in the blink of an eye.

Project / Auguste
Category / Home accessories
Design / Maisonnée
Photo / David Lemonnier

"MOTORCYCLE"

Motors are boys' best toy. What if your child is not old enough to ride one on the road? German designer Felix Götze turned a ratty bike into a "rocking bike" to fulfil all the little riders' dream. Built for a kid named Otto Komei, Ottos 150ccm is a mini-chopper in an old school bike appearance, with a high handlebar, classic parts, vintage lights and seat, fired by a one-cylinder two-stroke engine 150cc for good vibrations and sound with short exhaust, no muffler and skewed finish. Now he can learn how to keep his balance and sit cool chopper style.

Project / Ottos 150ccm
Category / Toy
Design / Felix Götze

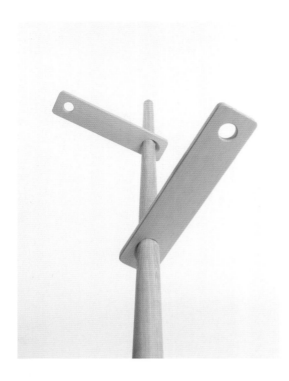

"MEASURE"

Growth is made to watch over children's growth. Just like a growth chart, this home furniture allows parents to measure their kids' height and adjust the hooks accordingly as they get taller and bigger and eventually catching up with the adults. The coat rack is produced from steel and beech.

Project / Growth
Category / Home accessories
Design / Ryosuke Akagi (Design Soil)

"MONSTER"

So this is true — your cupboard will blink and your drawers will stretch their hands and legs at night while you dozed off! Comprised of a cupboard 'Medusa', a chest of drawers 'Phil me in' and floating drawers 'The little guy', 'Puff&Flock' is the namesake London-based textile collective who collaborated with Stuart Melrose on this playful project. The prototypes are made from oak with lacquered surfaces to realise the coming to life of inanimate objects in a child's imagination.

Project / Puff&Flock
Category / Furniture prototype
Design / Stuart Melrose, Puff&Flock (Amelie Labarthe, Elisabeth Buecher, Melissa French)

Nn

"NEST"

Giant Birdsnest was conceived and created as a prototype to set new standards for socialising space. Cozy and informal, the wooden nest made of lacquered pine wood and filled with highly configurable egg-shaped poofs makes for a sensual refuge for family and friends to relax and interact. The nest's size and the colours of the poofs can be customised on demand.

Project / Giant Birdsnest
Category / Furniture prototype
Design / O*GE CreativeGroup

Oo

"OAK"

GROU is an adjustable coat rack which resembles a fruit tree in its broadest sense. The ripened fruits work as hooks which can slide along the rails to space out clothing items across the tree top. GROU is part of the "The forest at home" kid furniture collection which also features bookshelves, a rug and a chair.

Project / GROU
Category / Home accessories
Design / Menut
Client / Mirplay
Photo / Carlos Moreno, Carlos Lluna

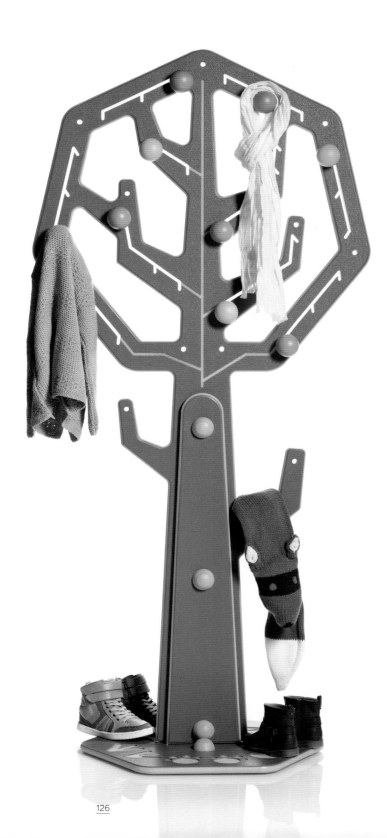

"OGRE"

Dreaming or daydreaming can sometimes be a great source of inspiration, and that's exactly where Joshua Ben Longo's eyeless monsters come from. With spiky hair, exposed teeth, open tummies releasing creepy worms, Longo's unique little creatures were said to exist in his midnight hallucinations with influences from childhood toys, b-rated sci-fi film, horror film, comics. Some of them even live in the artist's house surreptitiously as sculptures, rugs or chairs. Featured here are Monster Skin Rug[1], Monster Skin Chair[2], Little Whites[3], A Mighty Steed[4], Big Monster[5] made from felt, cotton, fur, leather and wood.

Project / Monster Sculptures
Category / Home accessories
Design / Joshua Ben Longo
Client / (Big Monster) Shelburne Museum

1

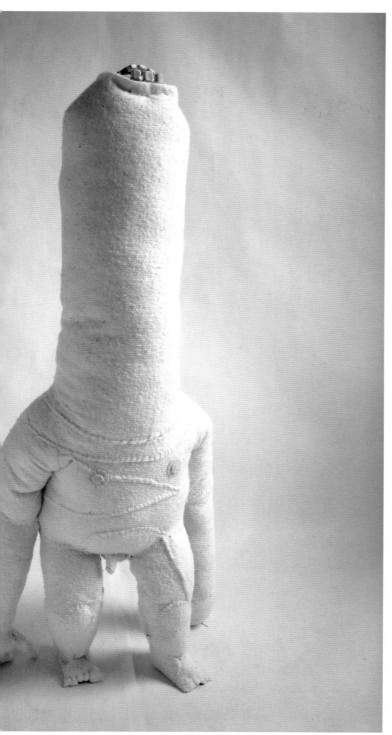

131

"OWL"

OWLANDER is a wish and mnemonic for all. While it represents wisdom and the remarkable ability to see in the dark, the 12 conical calendars, each of a different colour and character, also illustrate brotherhood as they stack up that essentially helps to get through difficulties. Each owl conveys a unique pattern which will come clear when it is lit inside.

Project / OWLANDAR
Category / Calendar
Design / RMM

Pp

"PAPERCRAFTS"

Originating as the legendary Japanese tale creature which the brand borrows for its name, TSUCHINOCO's playground equipments are equally fascinating in children's eyes. All made possible to be assembled by kids and their parents, the structures, singly built from reinforced corrugated fibreboard, reconstructs a rural village, believably to be the creature's favourable habitat. The line includes a tractor-slide, table sets, a rocking horse and a hut.

Project / TSUCHINOCO
Category / Playground equipment
Design / Masahiro Minami Design
Client / Nihon Logipack co., ltd.

"PORTRAIT"

Is this how you used to play with your face? Inspired by children's reaction towards their mirror images, especially with their permanent teeth growing out, Mine captures a bunch of happy kids being carried away in front of the mirror behind which Timothy Saccenti hid his camera for the photo shoot. Mine is an on-going project for an online fashion and fine art magazine focusing on the subject of childhood.

Project / Mine
Category / Photography
Photography / Timothy Saccenti
Creative direction / Alice Bertay
Production & Client / kid-in.net

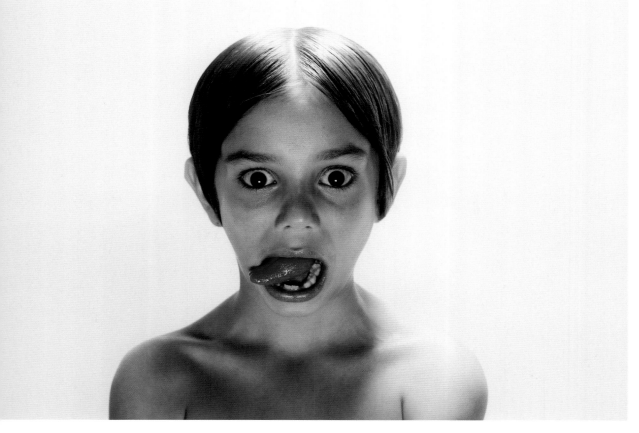

"PET"

A pet dog will surely light up your place, so will Frank. Taking inspiration from the delightful moments of having a cute puppy around in the house, Frank is trained to illuminate your room in the dark and stay right there as you command. Frank is conceived from beech wood with four hinged legs and a mouth that will always be wide open with a naked light bulb.

Project / Frank

Category / Lighting

Design / Pana Objects

"PUPPET"

My ROOTOTE says hi! Popping out from a tote bag's side, these simple but cute puppets are an extension of ROOTOTE bag's signature side pocket which are supposedly hidden like a kangaroo's pouch. Coming as a collaboration between nendo and ROOTOTE, Roopuppet features four characters — a bear, a dinosaur, a kangaroo and a guy. With its shape and depth expanded, the bag invites you to always keep your hand inside the bag to surprise.

Project / Roopuppet
Category / Accessories
Design / nendo
Client / ROOTOTE

"PUPPY"

Perhaps the most disastrous moment all dog owners know too well. Good Puppy and Good Boy are born out of a mischievous notion — both in a defecating pose with an On/Off switch in the shape of a turd that needs to be pressed or, in case of Good Boy lamp, stepped on. Both lamps were originally intended to participate in 'The Art Below' exhibition held at London Underground stations but eventually excluded for being too offensive.

Project / Good Puppy & Good Boy
Category / Lighting
Design / Whatshisname

"POLLUTION"

Horizon is an antidote to our gross overuse of non-renewable resources. Pouring gallons of chocolate syrup over children, figuratively like the real victims of the disastrous oil spills, this series makes for a powerful appeal that at once honour our past and prompt immediate action for the sake of the future through children's accusing gaze. The photo shoot took place at a bay to commemorate the thousands of gulls and pelicans whose life continues to be at stake.

Project / Horizon
Category / Photography
Photography / LaNola Kathleen Stone

"PARTY"

In a unique cooperation, Pictoplasma transformed a bevy of captivating, screwball and outstanding characters of our time into fantastic costumes. Padded, hydraulic or helium-filled, the creatures literally came to life and began to mingle with the local community of Berlin, Mexico City and TelAviv in the water and on the streets. Under the direction of choreographer Jared Gradinger, dancers and performers explored these new life forms and their individual character. Freed from the binds of storytelling and advertising, the characters developed their individual will, which they have proudly demonstrated in countless guerrilla style interventions during their ongoing tour.

Project / PictoOrphanage
Category / Costume design
Design / DOMA Collective, Pictoplasma
Photo / (Red creature) Daniela Kleint, (Blue creature) Miran Delija

"PLAYGROUND"

Dancing Houses are built to reflect the nature of the neighbourhood of Brumleby, one of Copenhagen's architectonic culture gems, as well as to turn everything topsy-turvy. Comprised of three houses, a baker and an ice-cream booth, the playground creates a piece of the old Brumleby and a wonder world for the little ones to run around. Climbing grips are all over the walls, with slides from the windows and balancing lanes from house to house. Children are free to rush in and out, and up and down.

Project / Dancing Houses
Category / Playground
Design / MONSTRUM
Client / Brumleby

"PLAYHOUSE"

LILIANE Dolls Villa could be a dream house for children, as well as for adults who have an eye for quality. Measuring 1.28 metres tall on metal castors with brake systems, the three-storey playhouse exhibits meticulous attention to details from its design through to the materials it employs. Open on four sides with perforated stainless steel roofs, the birch plywood structure reveals a kitchen counter and a catalogue of wood furniture next to functional wall mirror and aluminium utensils in seven rooms connected by two spiral staircases. The unblemished nature of wood has been consciously retained for a genuine connection between children and products of the earth.

Project / *LILIANE* Dolls Villa
Category / Toy
Design / *LILIANE* BV

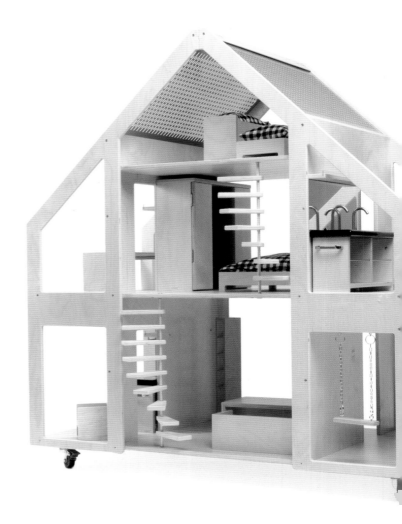

"PIGGYBACK"

Every child enjoys a ride on their father's shoulders. From there they can look afar like adults, hold their dad's head close and most importantly, save the walking! Named Abooba Chair, which literally means "piggyback chair" in Korean, the chair is dedicated to parents who occasionally wish to sit down to read without turning away from their kids. Incorporating a mini ropes course at the back of an armchair, the design gives children safe and easy access to the adult's shoulder and sit without transferring weight.

Project / Abooba Chair
Category / Furniture
Design / Jaewook Kim

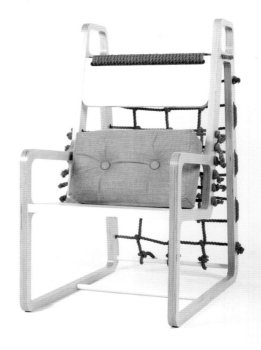

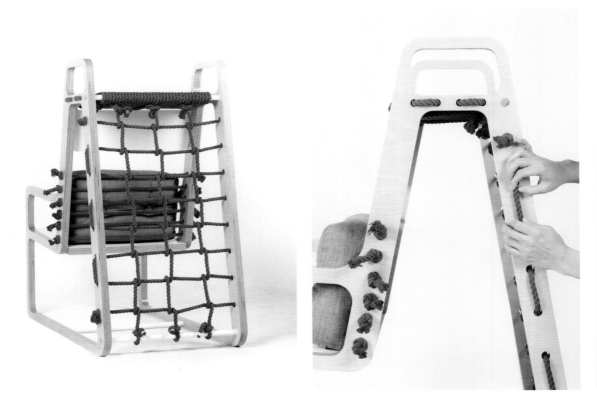

Qq

"QUEST"

Featuring an oblique desktop attached with seats of different levels, Growth Table introduces a new typology of utilitarianism that sponsors activities for multiple users at the same time. The zig-zag table top provides working space for three rows of seats to enter on alternative sides, and holes and indents to secure drawing tools and paper on its slanting surface. The table creates a community-scaled "art-parklet" when its form continues to expand to accommodate more users.

Project / Growth Table
Category / Furniture
Design / Tim Durfee & Iris Anna Regn
Photo / Jeremy Eichenbaum

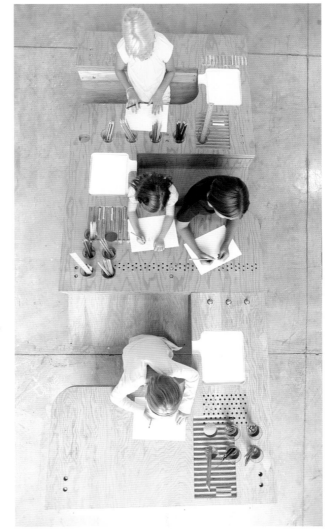

Rr

"RECEPTION"

Dublin advertising and marketing agency Boys and Girls understands how the reception area of an office sets the tone for a company and hence, the Balloon Reception Desk after commissioning a Lego table for their boardroom. The jolly-looking desk greets their existing and potential customers at their door with oversize Jenga blocks as the single leg support. The balloons and ribbons were reinforced to endure for as long as it takes.

Project / Ballon Reception Desk
Category / Furniture
Design / Boys and Girls
Manufacture / Twisted Image
Photo / Liam Murphy

"RESTAURANT"

Adults and children walking over to dinner at Phil can get themselves ready with separate plans. Opening its doors as a restaurant-cum-playground, the Bucharest diner welcomes families to eat next to a four metre-tall elephant and children to play downstairs. The ground floor features a playground and multi-purpose room which can accommodate activities from puppet shows to martial arts classes. Large colourfully-framed openings add dynamism to the space while allowing parents to keep an eye on their kids the entire time.

Project / Phil
Category / Interior design
Design / Nuca Studio
Client / Sorin Chirita
Photo / Cosmin Dragomir

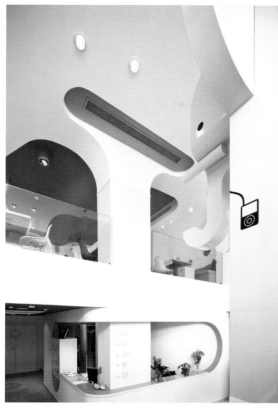

"REUNION"

It will not take long for anyone to get into the swing of things, whether it's a social gathering or boardroom meeting with staff returning from a long vacation. Available in bespoke finishes and sizes, Swing Table creates a room within a room with a faceted lampshade and matching chairs suspended from a solid steel frame. The four-poster table contains responsibly-sourced walnut and powder-coated mild steel.

Project / Swing Table
Category / Furniture
Design / Duffy London

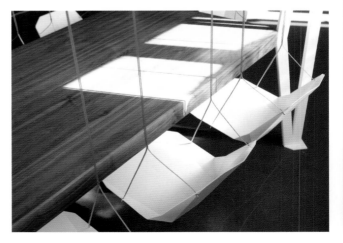

Ss

"SHOES"

Almost every girl admires the sexy looks they could possibly conceive from stiletto heels, Kobi Levi however sees a different quality in footwear art. Bringing a profusion of daily life elements into the silhouette of shoes, Levi elevates shoes to wearable sculptures that connect both children and women with a profound sense of humour. Over the spreads are a Sling Shot[1], Market[2], Slide[3], Dog[4], Tongue[5] and Market Trolley[6].

Project / Kobi Levi Footwear
Category / Footwear
Design / Kobi Levi Footwear Art
Photo / Shay Ben-Efraim, Ilit Azoulay

2

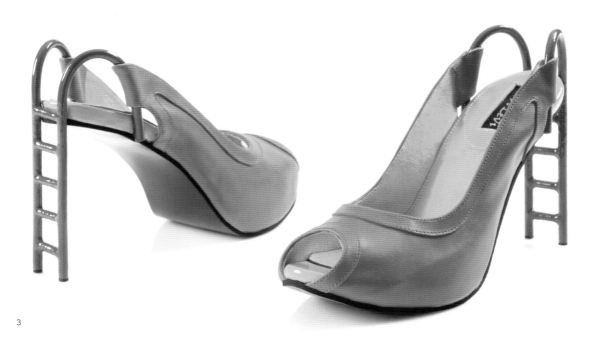

3

4

5

6

"STATUE"

In a way, Virgin Mary and Barbie are holy figures in two very different worlds. Little will identify them with one another, not until now. Through restoring and transforming damaged statues acquired in garage sales, Soasig Chamaillard cultivates a comical, and probably controversial, interaction between religious and global pop culture icons as a reflection of modern spirituality in today's society. Here presents Vierge de Couleur (Holy Colour)[1], Vierge au Petit Robot (Virgin Mary with Little Robot)[2], Hello Mary[3], Sainte Barbie (Saint Barbie)[4], My Little Mary[5] and Sainte Force (Saint Force)[6].

Project / Appearance
Category / Artwork
Design / Soasig Chamaillard

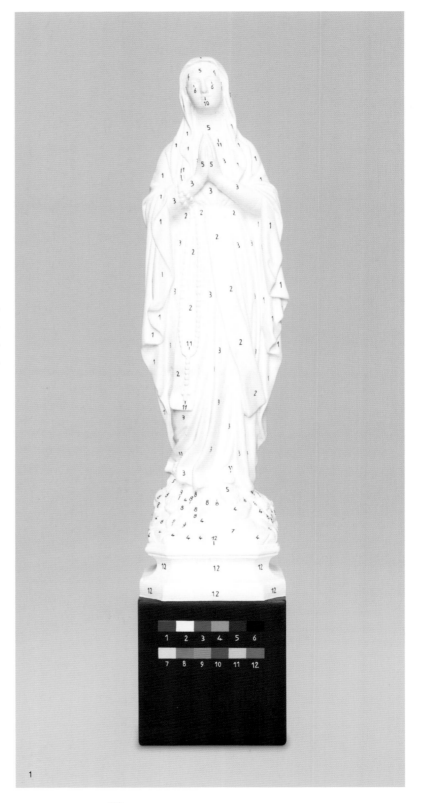

1

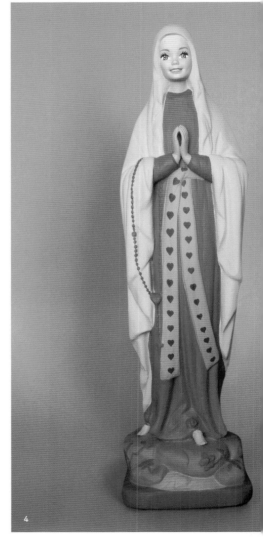

3

4

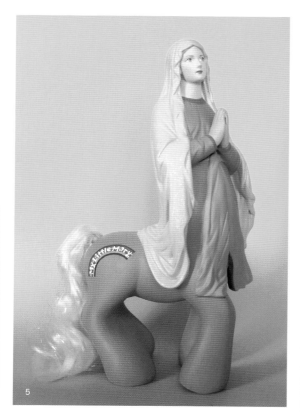

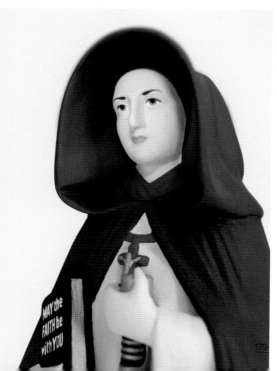

"SCRAWL"

Children's drawings depict a weird yet beautiful world with little sense of dimensions. But with 3D modelling technology, these semi-imaginative creatures will now have a chance to join our physical world with volume and weight. Prompted by a daughter's request for actualising a drawing of hers, Crayon Creatures is a service committed to turning children's doodles into tactile figurines using 3D print. Each product conceives a size of around 10cm in height, in full-colour sandstone, possible to be exhibited on a range of surfaces besides fridge doors.

Project / Crayon Creatures
Category / Artwork
Design / cunicode
Special credits / Lena Cuni Wong, Quim Cuni Wong

"SCARF"

Face changing can be easy as pie, with two mouths digitally printed on a silk scarf in a distinctive palette. "Happy/Sad" Scarf is a product of SWITCH! which offers a diversity of eyes, noses, mouths and haircuts drawn by FUNNY FUN With GUILLAUME on Dudes Factory, where users can customise designs for their own fashion accessories or T-shirt. The project pays tribute to Mr Potato Head, the clumsy, grim potato-shape guy who often finds himself falling apart in the Toy Story movies.

Project / "Happy/Sad" Scarf
Category / Art project
Design / FUNNY FUN With GUILLAUME
Client / Dudes Factory

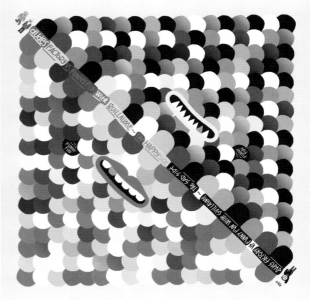

"SWAY"

What are the moments you relish most on a swing — flying up high, feeling the air passing as you move back and forth, or being pushed by your mum or dad from the back? With LED lighting incorporated into the seat, Swing Lamp conveys a beautiful ambiance where you can relax and gather memories without feeling alone in the dark.

Project / Swing Lamp
Category / Playground equipment, lighting
Design / BCXSY

"SWING"

Playground makes for an ideal gathering place for social interaction between children as well as adults and Rodrigo Caula is inspired to bring this scenario forward to the dining room. By removing the legs of the dining table and suspending the tabletop and seats, Eat & Play enlivens the space by putting everyone on the same swing. Let's break the ice with some swing fun.

Project / Eat & Play
Category / Furniture
Design / Rodrigo Caula

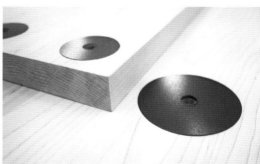

"SWEET"

Life is sweet — you can say that aloud by owning this rainbow-striped sugar-chair and letting it be your first step to complete your candy house. Fashioned out of almost 30 kilos of sugar, sugar-chair is a piece of hard candy which changes its shape and form as the user hugs it, sits on it or consumes it. Mold it however you like. But the more is consumed, the less mass will be left of the chair.

Project / sugarchair
Category / Furniture
Design / Pieter Brenner

"SEAT"

'EVA' is a durable, recyclable and non-toxic material integral to the making of this lightweight chair, specially designed for children. Cast out in one flat piece, the chair can be easily assembled by rolling it up and fastening the straps at the back and undone for storage. The chair comes in a range of vibrant colours that are most appealing to kids.

Project / EVA Chair for Kids
Category / Furniture
Design / h220430
Photo / Ikunori Yamamoto

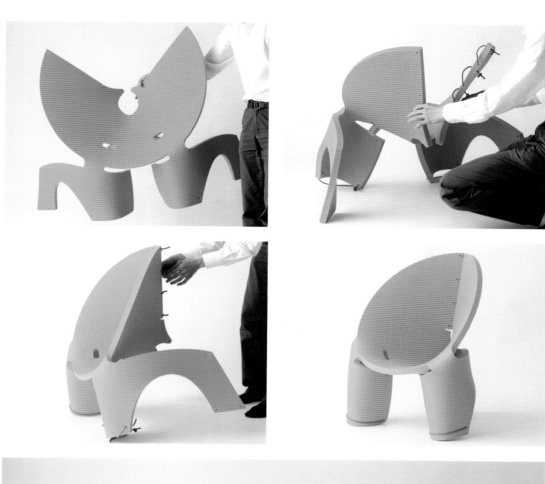
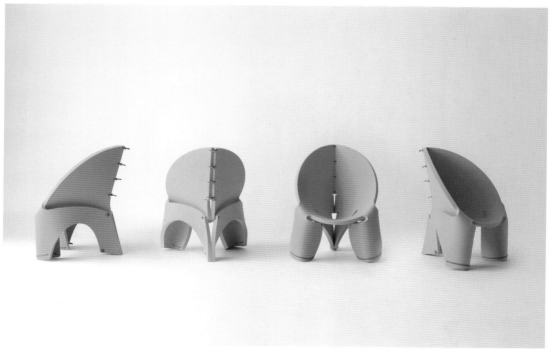

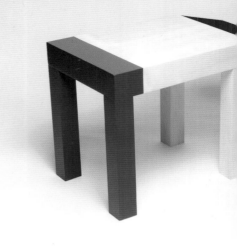

"SKETCH"

Animal reflects the unification of tables and dogs in young children's universe. Standing on four sturdy legs with a slanting compartment resembling a dog's head, the maple wood furniture takes on a simplified animal appearance based on the sketches made by a group of five-year-olds with convenient storage space for paper and painting utensils available within arm's reach. The table's surface has been set at a height which children can just knee and draw.

Project / Animal
Category / Furniture
Design / Quentin de Coster
Photo / Dominique Cerutti

"STORAGE"

Fun and learning can go hand in hand. Made to teach kids to find and sort their clothes, Training Dresser centres on graphics with drawers specially shaped to hint at its content. The cabinet comes in two designs, featuring clothing items ranging from tees to pants for boys and from sleeveless shirts to skirts for girls.

Project / Training Dresser
Category / Furniture
Design / Peter Bristol

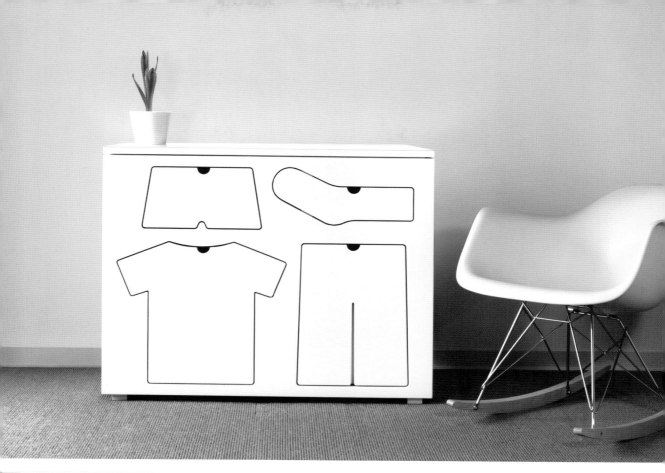

Tt

"TALK"

Kenno chairs introduce a new kind of recycled cardboard. Sturdy yet versatile and light in weight, the paper material sends out a sense of warmth and allow children to be creative with the chair's look. Each chair or stool is assembled with four pieces of cardboard without recourse to glue or screws. Its components, including water-based adhesives, allow it to be recycled when its little users outgrow the seats.

Project / Kenno Cardboard Chair
Category / Furniture
Design / Heikki Ruoho
Client / SHOWROOM Finland
Photo / Kalle Kataila
Model / (Boy) Noah

"THEATRE"

THEA chair is a hybrid of theatre and chair. Where raised grand drapes are permanently cast at the back of the chair, the chair naturally becomes an acting space where kids can improvise plays with siblings or friends any time. Underneath the chair is a secret compartment where finger puppets can be stored. The lid, once opened and locked, conceals the backside of the chair and allows some privacy for the puppeteers.

Project / THEA chair
Category / Furniture, toy
Design / Menut
Client / Mirplay
Photo / Carlos Moreno, Carlos Lluna

"TOOTH"

Asked to create a space to ease children's stress while waiting for their dental examination, TERADADESIGN decided to erect a tranquil zone populated with woods, life-size animal decals and curious worms in Tokyo. Besides a harmonious blend of colour and materials, the interior also incorporates games and structures for kids to explore. A thoughtful glass partition allows parents to remain in sight during the entire consultation process.

Project / Matsumoto Pediatric Dental Clinic

Category / Interior design

Design / TERADADESIGN

Client / Matsumoto Pediatric Dental Clinic

Photo / Yuki OMORI

"TONGUE"

Used to be a horror story concocted by the designers' granny to stop him from playing with fish balls during his meals, the mythical cannibal fish ball came real as a gigantic hairy spherical monster Jason Goh envisioned whenever he was warned. A real man-eater but definitely not scary in every aspect, "Moyee" embraces you with its warm and cozy body. Like a roly-poly toy, Moyee allows its friends to rock inside without rolling over.

Project / Moyee
Category / Furniture
Design / Jason Goh
Special credits / Little Thoughts Group

"THREESOME"

Never feel lonely again watching TV by yourself on a Sunday evening. Sofa so Good is your ideal companion with long welcoming arms reaching out to soothe you as long as you sit in 'him'. Befriend 'him' by brushing and styling 'his' hair-cushions when you have guests around. Exhibiting excellent craftsmanship, the comfortable three-seater is a collaboration between Puff&Flock and Ercol, a traditional furniture company based in England.

Project / Sofa So Good
Category / Furniture
Design / Puff&Flock (Elisabeth Buecher, Amelie Labarthe)
Client / Designersblock, Ercol

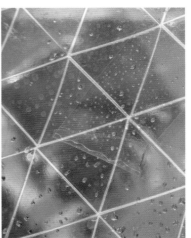

Uu

"UMBRELLA"

Rain might interrupt games and picnic plans, but there might as well be merry moments if you have a cheerful soul like Gene Kelly. Draw a rainbow with these umbrellas before the rain stops. Appearing to be grey at first sight, the canopy composed of single-sided deflection films produces rainbow-like patterns when two of the kind overlap. The colour continues to transform as you twirl the umbrellas.

Project / Umbrellas of Rainbow
Category / Accessories
Design / takram design engineering, KOKUYO FURNITURE Co., Ltd., Panasonic Electric Works Co., Ltd., FUJITSU DESIGN LIMITED
Photo / Takashi Mochizuki

Vv

"VARIETY"

Ping Pong is a transformation of an ordinary everyday object into something more. Conceived in the size of a coffee table with a chalkboard surface, this free standing furniture sets off living room as a convivial counter to hold your drinks, ping pong table, a desk and a message board. Ping Pong comes along with matching rackets, ping pong balls, chalk and eraser. The simple pleasure doubles up with ecological packaging.

Project / Ping Pong
Category / Furniture, toy
Design / Mike Mak
Client / Huzi Design

Ww

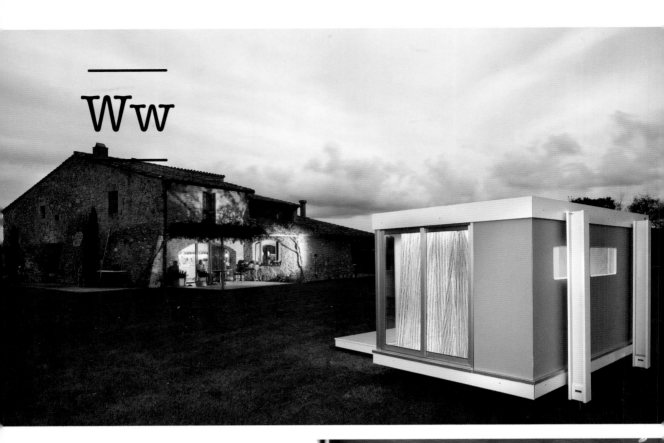

"WENDY HOUSE"

There will be one more way for children to wallow in great architecture besides visiting a house museum or reading books. Taking reference from the brilliance of architecture conceived in the 20th and 21st century, SmartPlayhouse invites children to explore some genius' frame of mind, like of Ludwig Mies van der Rohe when he built the Farnsworth House and perhaps also of Toyo Ito for MIKIMOTO Ginza 2. The houses were meant for outdoor use.

Project / SmartPlayhouse
Category / Playhouse
Design / David Lamolla Kristiansen (SmartPlayhouse)
Photo / Eros Albarrán, STARP ESTUDI

"WEB"

Takino Rainbow Nest and Hakone Forest Net are two of textile artist, Toshiko Horiuchi MacAdam's NetPlayWorks project entirely crocheted by hand. Sturdy, flexible and resilient, these vivid structures lure children to crawl in, roll around and jump in a fanciful version of our adorable environment. Hakone Forest Net was installed in Woods of Net, a permanent pavilion built in collaboration with T.I.S. & PARTNERS at Hakone Open Air Museum, Hakone, Japan.

Project / NetPlayWorks
Category / Playground
Design & Construction / Toshiko Horiuchi MacAdam, Interplay Design & Manufacturing, Inc.
Structural design / T.I.S. & Partners
Project design / Takano Landscape Planning
Photo / Masaki Koizumi

Xx

"XYLOPHONE"

Like an oversize xylophone, the new annex to Kindergarten Kekec in Ljubljana, Slovenia is characterised by large windows and vertical shutters along all of its three exterior walls. Painted on one side with its natural colour retained on the other, the vertical shutters reveal a rainbow in nine bright colours as they are spun open to let daylight in. The kindergarten's appearance evolves as the timber slates rotates at different point of the day.

Project / Kindergarten Kekec
Category / Architecture
Design / Arhitektura Jure Kotnik
Structural engineering / CBD d.o.o.
Mechanical engineering / Linasi Peter
Client / Mestna občina Ljubljana
Photo / Miran Kambič

Yy

"YACHT"

A huge cargo ship is spotted adrfit next to a lighthouse, leaving everything in a state of utter chaos! Sending children off for a treasure hunt out of harm's way, the playground is designed to fit into the maritime environment around the grass area at the marina, as part of Höganäs municipality's renovation scheme. The kids are challenged to move on a dangerous route between loose boxes and driftwood while keeping a balance on debris in the hull for then flee to safety at the lighthouse, away from danger at the ocean.

Project / The Cargoship in Höganäs
Category / Playground
Design / MONSTRUM
Client / City of Höganäs

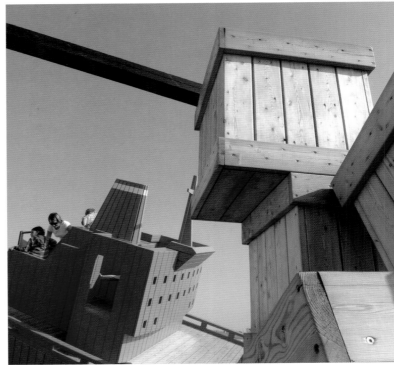

Zz

"ZONE"

Cargo containers do not seem to be a comfortable setting for schooling, but in Kindergarten Ajda's case, sea cans have proved to be an excellent and flexible solution. Originally a short-term solution for the immediate relief of insufficient kindergarten space (on this spread), the idea was expanded to become its current permanent structure (on next spread) as an approved didactic site, housing three classrooms, two covered terraces and two washrooms for children, fluidly connected with the dressing room and a multi-purpose entrance hall. The school has a magnetic façade made from thick anthracite isolative and fire-resistant boards, covered in colourful magnets that can be modified by children at ease.

Project / Kindergarten AJDA 1 & 2
Category / Architecture
Design / Arhitektura Jure Kotnik
Container system / Conhouse
Client / Municipality Ravne na Koroškem
Photo / Miran Kambič

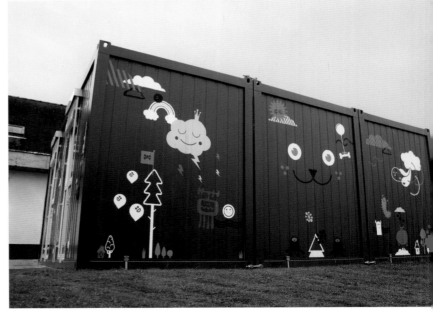

Biography

Akkoyun, Ozan

Taking engineering initially, Akkoyun graduated from Mimar Sinan Fine Arts University in graphic design, He is currently working as an art director at Hep and lives in Istanbul.

P. 086-087

Arhitektura Jure Kotnik

Based in both Ljubljana and Paris, Jure Kotnik works in various fields of architecture from research to design. He has designed several architecture realisations that brought about media attention such as Weekend house 2+ and Mobile Lighthouse for Paris Port. He is also the author of the best-selling monograph *Container Architecture* and the curator of the first-ever container architecture exhibition in Ljubljana, Paris, Berlin and Seattle in 2010. He has dedicated himself to kindergarten architecture in the past three years and designed several contemporary kindergartens in Slovenia. He was appointed the visiting professor at Ecole Speciale d'Architecture in Paris in 2012 Autumn.

P. 222-225, 228-231

BCXSY

Established in 2007, BCXSY is a cooperative between designers Boaz Cohen and Sayaka Yamamoto. Besides working on their own projects, they also deliver multi-disciplinary services on identities, products, graphics and interiors. BCXSY's work has been featured in some of the world's most prestigious events and is currently part of the permanent collection at the Victoria and Albert Museum.

P. 096-097, 186-187

biaugust CREATION OFFICE

In 2005, Owen Chuang and Cloud Lu founded biaugust CREATION OFFICE in Taipei. The duo insists on infusing "life and emotion" into their creations and explores the possibilities between human and design with a variety of media. They have both received numerous design awards from Japan, Hong Kong and Taiwan.

P. 030-031, 114-115

Bob Foundation

Bob Foundation was established by two graduates from Central Saint Martins College of Art & Design, Mitsunori Asakura and Hiromi Asakura in 2002. They actively collaborate with overseas artists and designers on various artwork and also produce furniture, kitchenware, bags and wrapping papers under their name. They launch a new paper brand "Number 62" in 2007.

P. 018-019

Boys and Girls

Boys and Girls is a creative agency based in Dublin, specialising in advertising, marketing and digital communications with occasional adventures into furniture-making. Set up with the simple idea of "Great work, works" in 2009, the agency works for a diverse array of clients from telecommunication companies, car manufacturers to the alcoholic beverages industry.

P. 164-165

Brenner, Pieter

Born in 1976, Brenner started to make his own products and designs in 2010. His work manipulates the thin line between dream and reality and is not limited to one style, method or identity. The designer and artist lives and works in Cologne, Munich, New York and Buenos Aires.

P. 190-191

Bristol, Peter

Working to create, evolve, and refine ideas and brings them to life, Bristol is a prolific industrial designer that has delivered groundbreaking and award-winning solutions across genres. His work often creates new architecture that becomes inherently recognisable products. He is currently the Creative Director of Carbon Design Group, a Seattle-based product development firm.

P. 196-197

Caula, Rodrigo

Graduating from Emily Carr University with a Bachelor of Design, Caula is an industrial designer based in Milan, Italy.

P. 188-189

Chamaillard, Soasig

Born in France in 1976, Chamaillard works with damaged statues found in garage sales. The playful interaction of these statues through restoration and combination serves as a reflection on Chamaillard's perspective on life as well as women's role and place in the society.

P. 176-179

Biography

Christian Vivanco Design Studio

Vivanco has been characterised for his strive for perfection in his work. His collaboration with organisations such as Hewlett-Packard and Santa & Cole has created celebrated products throughout the world.

P. 029

cunicode

The design studio established by Bernat Cuni is keen to explore scenarios of creation, production and consumption through design, digital fabrication and 3D printing.

P. 180-183

de Coster, Quentin

Born in Liège in 1990, de Coster is a young Belgian designer who studied industrial design at the ESA Saint-Luc Liège and graduated at Politecnico di Milano in 2011. Having great interests on manufacturing processes, de Coster works on projects ranging from products to furniture with international partners.

P. 194-195

Design Soil

A design project at Kobe Design University aiming to explore on experimental themes in today's educational context. By re-evaluating issues like the way of thinking, manufacture process and marketing, students are asked to challenge the common perception of ordinary in existing designs. It's a place for young designers to think and create things one by one, slowly yet steadily like plants grow in soil.

P. 120-121

Design Systems Ltd

Established in 1999, the Hong Kong-based interior and industrial design studio provides consultancy services on corporate offices, branding and commercial retails, public spaces and urban furniture, exhibition, signage and art installation design. The majority of its projects completed in the past 5 years have been honoured nearly 100 reputable international design awards.

P. 088-089

Diego Ramos Studio

Based in Barcelona, the design studio has been collaborating with creatives and clients around the world since 2005. Born in 1978, Ramos studied at Eina - Escola de Disseny de Barcelona and graduated with a master's degree in design product at the Royal College of Art, London. Ramos also teaches workshops in design universities and develops personal projects that try to explore everyday live and question the role of the design in today's "product environment".

P. 104-107

DOMA Collective

DOMA Collective is set up by a group of Argentines artists who started in the Buenos Aires Street-Art scene back in 1998. The group specialises in animation, filming, illustration, installation art, scale models, objects and toy design. They have been invited to contribute at various international projects in Berlin, New York, São Pablo, London, Barcelona, Montreal, Israel and so on. Group members include Mariano Barbieri, Julian Pablo Manzelli, Matias Vigliano, and Orilo Blandini.

P. 152-155

Duffy London

Founded by Chris Duffy who graduated at the University of Brighton, Duffy London works to produce furniture, lighting and interior products with new production direction and methods in a environmentally-friendly way.

P. 082-085, 170-171

Durfee, Tim and Regn, Iris Anna

Tim Durfee, an expert on conceptual design and media and Iris Anna Regn, who excels in social practice and material experimentation teamed up to produce award-winning exhibitions and installations for major cultural institutions, small buildings and urban communication systems. Durfee also conducts his design research through amp, a studio he operates through Media Design Practices at Art Center College of Design, while Regn is also a co-founder of the ongoing art and design project BROODWORK: Creative Practice and Family Life. They are based in Los Angeles, California, USA.

P. 162-163

DvanDirk

Creating and developing products with two fundamental characteristics — natural and transparent, DvanDirk designs without hiding the way the product works or are constructed, but instead showcasing them with justifiable pridem, making durable products which spare the environment.

P. 028

Bb

Biography

Front

Set up by Sofia Lagerkvist, Charlotte von der Lancken and Anna Lindgren, the Swedish design group's objects often communicate a story to observers about the design process, the material they are made of and the conventions within the design field.

P. 014-015

FUNNY FUN with GUILLAUME

FUNNY FUN with GUILLAUME's bold and colourful universe evolves in different fields such as illustration, graphic design, installation and street art. In the society overdosed by images and information, his work stands out by its simplicity and coolness.

P. 184-185

Goh, Jason

A Singaporean designer graduated with a degree in design from the University of Glasgow in Scotland in 2001. He started out as an industrial designer at GE/Fitch and has since worked in several multi-national companies including Panasonic, Siemens VDO, Creative Technology and Hewlett-Packard on a wide variety of design projects such as audio/video hi-fi systems, LCD TVs, car audio, gaming machines, multi-media speakers, mobile phones and business printers.

P. 204-205

Götze, Felix

Born in 1986, Götze graduated with a master's degree in industrial design at University of Art and Design Halle, Germany in 2012. He was an architecture intern at GRAFT Berlin, a product design intern at GRAFT Los Angeles and was also on an internship at Deisig-Design Berlin during his studies.

P. 118-119

Grauer, Felix

Grauer is an architecture student at Cologne Institute of Architectural Design. Upon completing his bachelor degree in Regensburg, he moved to Cologne to do a master's degree on coporate architecture. During his studies, Grauer worked in several architectural offices. He founded a study community, Äbsolut Office, to take on architecture and urban art projects with his university friends. Grauer was also a tutor and contributed to a research project in his university.

P. 010

Guillaumit

An illustrator, graphic designer and motion designer from France, Guillaumit works to build a universe both ludic and meaningful with geometric forms, rigid colours and funny cartoon characters. He also works in Gangpol & Mit, making videos that were played in concerts held in countries like Germany, Spain, Belgium, Mexico, Argentina, Japan and Taiwan.

P. 046-049

h220430

Established on 30th April, Heisei 22 (the name of the current era in Japan), h220430, commemorates this date by its name. They focus on lighting and furniture that communicate and provide opportunities for people to rethink and act against topics of everyone's concern such as deterioration of the global environment and continuous conflicts all over the world.

P. 098-101, 192-193

haoshi Design

Adopting the Chinese pronunciation of "good things" in their name, the team believes that life is full of happiness and interesting things. Established in 2009, haoshi Design works to bring people peace, joyfulness and satisfaction through concrete and exquisite technical visual art, where the abstract concepts of life and the designers' philosophy are presented.

P. 016-017

Héctor Serrano

A design office founded by Héctor Serrano in London in 2000. The studio's clients include companies such as FontanaArte, Roca, ICEX Spanish Ministry of Industry, Tourism and Trade, Droog and Metalarte among others. The team has received the Peugeot Design Award, the Premio Nacional de Diseño No Aburridos and the second prize of the New Bus for London competition. Their products have been exhibited extensively in museums such as the V&A in London, the Cooper-Hewitt National Design Museum in New York and the Central Museum of Amsterdam. Héctor studied industrial design in Valencia before moving to London to study a master's degree in product design at The Royal College of Art.

P. 008-009, 056-057

Biography

Huzi Design

A collection of smart and playful design objects that keep the curious sparkles among kids and adults. By telling meaningful stories, these smart objects invite audience to touch, play, laugh, love, think and learn. Huzi works with talented designers from all around the world who share the same vision to think beyond the ordinary. Handcrafted with unyielding attention to detail and quality, Huzi's products are timeless collectables that are used, loved, played with, and passed on from generation to generation.

P. 210-213

Imamura, Hikaru

Graduated in 2003 from the Visual Communication Design Department at Musashino Art University, Japan, Imamura worked at Kei Miyazaki Design in 2007 and later for Man and Activity, Product Design Department of Design Academy Eindhoven, the Netherlands in 2011.

P. 066-069

Jeong, Yeongkeun

An industrial designer based in Seoul, Jeong works mainly on electronics, packaging, stationery, artistic production and installation. His work is analytical and based on research, striving to solve problems that people experience daily while emphasising on a clean and refined look. Jeong has recently gained industry-wide recognition after he was awarded the Core77 Design Awards in 2011 and has been featured in many design sites and blogs, such as Behance.net, Lovely Package and Dieline.

P. 044-045

Jullien, Jean

A French graphic designer living and working in London. Jullien comes from Nantes and did a graphic design degree in Quimper. He graduated from Central Saint Martins in 2008 and from the Royal College of Art in 2010. He works closely with the musician Niwouinwouin on projects ranging from illustration, photography, video, costumes, installations, books, posters to clothing. In 2011, Jullien founded Jullien Brothers, a duo specialising in moving images. He created News of the Times with Yann Le Bec and Gwendal Le Bec the next year.

P. 108-111

Kidsonroof

A permanent testlab established in 2005 by Romy Boesveldt and Ilya Yashkin, together with their three kids and their little friends. Yashkin is an architectural designer trained at Marchi in Moscow and Rietveld Academy in Amsterdam while Boesveldt specialises on conceptual development and styling.

P. 070-071

Kim, Jaewook

Having studied interior design at Kookmin University, Kim has worked with the exhibition team for the Green Garden Panorama of Seoul Design Festival in Mapo-gu, South Korea as well as his graduation exhibtion in the university.

P. 160-161

Kobi Levi Footwear Art

Having graduated from Bezalel Academy of Arts & Design, Jerusalem in 2001, Levi specialises on the design and development of footwear. His creations transfer the essence of daily objects into the shape of shoes. Most of his inspirations are drawn from outside the "shoe world", giving the footwear an extreme transformation that is humorous with a unique point of view. Levi's creations were first uploaded to a blog he opened in 2010, which soon took up life of its own with over-night exposure that brought Lady Gaga to use the design Double Boot in her "Born this Way" music video. The shoes are exhibited around the world: the SONS (shoes or no shoes) museum in Belgium and "Going Bananas" group exhibition in Switzerland for instance. The handmade leathers shoes are sold in limited editions of 20 pairs per design.

P. 172-175

Kukkia

Meaning "to bloom" or "flowers" in Finnish and with the word "kuki" referring to stem in Japanese, Kukkia was launched in 2009 by Kazuyo Shiomi and Nobuko Shigeoka as a design production company creating design objects and toys from natural FSC responsibly sourced sustainable wood. Aiming to bring smiles to its audience, the company has developed their own brands, gg* and kiko+, while also collaborating with international designers. Kukkia also deals with Original Equipment Manufacturing (OEM) products for big companies. VIEKO DIS HK is the exclusive distributor in Asia except Japan.

P. 058-059, 072-075

Biography

Licata, Anna

Born in Turin in 1978 and having graduated in architecture at Polytechnic of Turin in Italy in 2003, Licata works mainly on public buildings, residential, commercial and office buildings while also developing projects with a special focus on children.

P. 036-037

LILIANE BV

A sustainable-living furniture brand for dolls up to 30 cm, LILIANE BV's products are built not only for fun but also with educational purpose. Targeting the three to ten age group, they create interaction that stimulates spontaneous practices of social skills. The use of materials also brings in the concept of sustainability for the kids at young age. Its value is as well-recognised in the field of child as women psychotherapy in the Netherlands.

P. 158-159

Lin, Yu-Nung

The Taiwanese designer is currently studying child culture design in Sweden. With a diverse range of creative experience and knowledge, Lin creates playful furniture and products for children and every age group.

P. 038-039

Little Red Stuga

The team innovates to support children's fundamental needs and human rights to engage in activity and be stimulated. They specialise in creating stories, products and projects with international appeal and pedagogical content for children and young-at-heart adults.

P. 112-113

Longo, Joshua Ben

Longo is an artist and a teacher. He is available for lectures, workshops, private commissions and/or commercial work.

P. 128-133

MacAdam, Toshiko Horiuchi

Born in 1940, MacAdam is one of Japan's leading fibre artists who uses knitting or crocheting in their work. She attended Hibiya High School and studied fine art at the Tama Art University, Tokyo, followed by a master's degree at Cranbrook Academy of Art, Michigan. Living in Canada, MacAdam now specialises in creating large, interactive textile environments that function as imaginative and vibrant explorations of colour and form, as well as providing thrilling play environments.

P. 218-221

Magini Design Studio

Raised in Arezzo, Italy and graduating from Milano Politecnico, Emanuele Magini was a set designer for Walt Disney Italy and an instructor for private institutions in the field of 3D computer modelling. His projects have won national and international awards such as Good Design Award from the Museum of Architecture and Design in Chicago, iF Product Design Award, and the ADI Design Index. Magini's clients include Heineken, Campeggi and Seletti.

P. 060-061

Maisonnée

Founded in 2012 by Audrey Belin, the design studio specialises in home decor objects. After studying fashion at the École Duperré and having obtained a master's degree in object design from the École Nationale Supérieure des Arts Décoratifs in Paris, Belin has worked alongside with her assistant Inga Sempé for international clients like L'Oréal, Bacardi and Areva. Belin's creations transform daily life to a fairy tale, requiring individuals to rediscover the form and utility of the object.

P. 053, 116-117

Masahiro Minami Design

Born in Osaka in 1978, Masahiro Minami graduated with a master's degree in environmental planning, design and architecture from the University of Shiga Prefecture. The frequent international award winner was a university research assistant in the Department of Living Design before establishing his own studio in 2008.

P. 064-065, 136-137

Matteo Bianchi Studio

The studio of the leading international interior designer who is based in London. Bianchi was born in Venice and studied at the University of the Arts, Chelsea, London. He set up the studio in 2006 and has since worked with international clients in Italy, the Middle East, Australia and the UK. Besides working on both commercial and residential projects, Bianchi also designs a range of products under the name of Daruma Design, the debut of which has been featured in international publications such as Vogue Italia and Zurich Deluxe.

P. 102-103

Bb
Biography

Megawing
Megawing International Group Co.,Ltd was established in 2008. It professionally exports products and accessories that are mainly designed and produced by Taiwan design team "Afterain Design" and other famous brands including CiCHi, Hsu-creation, Koan+ and Yedou.
www.megawing.com.tw

P. 022-023

Melrose, Stuart
Mixing inspiration from popular urban culture and 21st-century technology, Melrose embraces avant garde materials and production techniques, pushing the boundaries of what is possible in the arena of fine furniture.

P. 122-123

Menut
Based in Valencia, Spain, the design studio focuses on children design with an innovative vision on the way of life producing interactive and humorous designs that are timeless and interesting to people at all ages.

P. 032-033, 126-127, 200-201

miller goodman
The collaboration between Zoe Miller and David Goodman is based in Brighton, the UK. The duo works to create modern yet timeless designs that are inspirational, artistic and ecological for both kids and adults. Besides working for the advertising, fashion and creative industries, the team has also enjoyed an exciting collaboration with The Tate, producing four children books and associated product ranges.

P. 020-021

Monkey Business
Founded by industrial designer Oded Friedland, Monkey Business provides fresh perspectives on little things that make up your day, whether at home, at the office or outdoors, adding something extra to the ordinary.

P. 034-035

MONSTRUM
Specialising in designing and manufacturing unique playgrounds with a focus on artistic, design and architectural quality, MONSTRUM works to captivate and inspire both adults and children, tickling their curiosity with stories and details.

P. 156-157, 226-227

Nakanishi, Kana
Born in Germany and raised in the states and Japan, Nakanishi graduated from Keio University, Japan and worked for Kokuyo Co., Ltd. as a product designer before moving to Finland to further study about design at Aalto University. Nakanishi is now back in Tokyo and works as a product and interior designer.

P. 094-095

nendo
Founded by architect Oki Sato in Tokyo in 2002, nendo upholds its goal of bringing small surprises to people through multi-disciplinary practices of different media including architecture, interiors, furniture, industrial products and graphic design.

P. 144-145

Nuca Studio
Nuca works in multi-disciplinary approaches and believes that complex thinking leads to simple lines, it is all about details and the time spent designing a building is always less than 0.5% of that building's life.

P. 166-169

O*GE CreativeGroup
Founded by award-winning architects Merav Eitan and Gaston Zahr who specialise in architecture, urban and branding projects with highly creative, inspiring character. It is the interaction of different projects which keep O*GE's work fresh and pushing boundaries while being responsible within the framework of budgets and time schedules.

P. 124-125

Oscar Diaz Studio
Diaz studied industrial design at the Ecole de Beaux-Arts of Bordeaux and later got a master's degree in product design at the Royal College of Art in 2006. Much of his work is concerned with twisting everyday objects or situations by mixing influences from craft and technology. His approach is multi-faceted, investigative and playful, transforming the ordinary by challenging conventions. Diaz was born in Spain and currently based in London.

P. 011, 050-051

Paluchová, Veronika
Having obtained a master's degree from AFAD in Bratislava, Paluchová works as an interior and product designer creating things for everyday life with humour and individual insight.

P. 024

Bb

Biography

Pana Objects

Based in Thailand, Pana Objects is the brainchild of a group of designers and makers. With its aim to reintroduce the subject of woodworking and craftsmanship back into modern life, Pana Objects provides a range of thoughtfully-designed objects including household items, decorative product, stationery and many more.

P. 142-143

Pani Jurek, Gang Design

Pani Jurek and Gang Design belong to a group of designers from Poland who focus on the innovative use of materials and sustainable design. The brand was established by Magda Jurek, a Warsaw-based designer who graduated in painting from the Warsaw Academy of Fine Arts in 2007. Jurek's designs explore new ways of looking at the interaction between the object and the user. Lighting is a particular focus, along with socially-driven projects in the public space. She set up the Association "Based in Warsaw" with Edyta Oldak which won the Mayor of Warsaw's Award as the Best Extra-governmental Initiative in 2010 for Great Architecture for all Children. Jurek now lives and works in Ignacow near Warsaw.

P. 090-093

Pereira, Rui and Fukusada, Ryosuke

Joint forces of a Japanese and a Portuguese who came to Italy with the same goal — to become a product designer. The duo met by chance in a design studio in Milan and their interest to share their cultural background gave birth to a new tradition.

P. 027

Pistacchi Design

Established in 2011 by Mike He and Henry KH Huang. "Pistacchi" refers to pistachio in Italian, and is interpreted as "happy nut" in Mandarin, which is what the duo aims for: to deliver happiness and new perspectives. Most of their designs are drawn from daily life, integrated with stereotypes and experiences, but expressed from a new perspective so that users can be surprised by their own interpretations.

P. 054-055

Poulain, Damien

The graphic designer and art director has been working on a wide range of fields but more specifically on art, fashion and music projects in London since 2002. Poulain is also the founder of the publishing house, oodee, which books have been recognised by the photography community for their outstanding design and art direction.

P. 078-081

Puff&Flock

Puff&Flock is a collective of six textile designers united by their bold values and risk-taking approach. Formed in 2008 after meeting at the Central Saint Martins when obtaining master's degrees on textile futures, the collective debuted their eclectic spirit to much acclaim at Interiors Birmingham in 2009 with the support of Designersblock. Since then, they have gone on to entertain audiences at exhibitions in New York, London, Milan, and Birmingham with a variety of ambitious textile installations.

P. 122-123, 206-207

PUTPUT

PUTPUT is a Swiss-Danish artist group established in 2011 and currently based in Copenhagen, Denmark. Seeking to occupy the space between input and output, PUTPUT works in the field of contemporary art photography, sculpture, installation and publishing.

P. 076-077

Reddish

Formed by Naama Steinbock and Idan Friedland in 2002, the design studio produces a range of products from furniture, lighting and home accessories to jewellery characterised by clear shapes and simple structures.

P. 034-035

RMM

Also known as Readymade Creative, RMM is founded by Raymond Man Chung Lee and Amber Fu who both worked at IdN magazine and CREAM. In 2005, the duo co-founded the lifestyle magazine, Readymade Magazine.

P. 052, 134-135

Ruoho, Heikki

Born in 1969, Ruoho is an industrial and furniture designer graduated with a bachelor degree from the Lahti Institute of Design and a master's degree from the University of Art and Design Helsinki. Ruoho and Teemu Järvi set up the Järvi & Ruoho design office in 2003. Ruoho has received a number of national and international awards such as the Ornamo Design Award in 2008 and the Fennia Prize in 2007.

P. 198-199

Bb
Biography

Saccenti, Timothy

The director and photographer lives in New York City and works worldwide. Saccenti's singularly unique, immersive style has been imprinted onto genres of music, publication and fine art. Saccenti's groundbreaking visual output includes influential and genre-defying films for artists ranging from indie icons Animal Collective, hip-hop master EL-P to legends Depeche Mode, Usher and Erykah Badu. His commercial oeuvre includes the controversial "Playfaces" television campaign for the launch of the Sony PS3, as well as advertisements for Nike, Diesel and Audi amongst others.

P. 138-141

SmartPlayhouse

SmartPlayhouse is founded by David Lamolla Kristiansen in 2009, specialising in children`s playhouses. Studied architecture at the Politechnical University of Catalunya, Lamolla has worked with famous Belgian architect Mario Garzanitti and the hotel chain El Bulli Hotel. He was also the Co-director of the architectural firm ToolStudio SL for four years.

P. 214-217

SNURK

A small independent label based in Amsterdam, the Netherlands, SNURK designs exclusive bedding which are sold worldwide. it offers quality bedding and cushions with photographic print, designed by Peggy van Neer & Erik van Loo.

P. 040-043

Stone, LaNola Kathleen

A New York City-based photographer, artist, author and educator, Stone's commercial clients seek her out to photograph interiors, portraits and lifestyle although she is mostly know for her aptitude with children and childhood imagery. In 2012, Focal Press published her first book, *"Photographing Childhood: the image & the memory"*. Stone is also a second-class Petty Officer/Public Affairs Specialist (photojournalist/media liaison) for the United States Coast Guard Reserve.

P. 148-151

Suck UK

Offering products that are fun, exciting, unique with a sense of humour, leading British gift brand Suck UK is established by design duo Sam and Jude in 1999 after studying design at Central Saint Martins. The reputation of their original and quirky products has earned them multi-awards, bespoke projects and fast growing success with the opening of their flagship stores located in London's OXO Tower and Westfield at Shepherd's Bush.

P. 026

takram design engineering

The author of *"Storyweaving"* on synchronicity of ideological structure and development process, Takram, aka Kotaro Watanabe, participates in diverse designs of digital user interfaces and interactive installations. Delegated as a resident at Stockholm by Swedish Arts Grants Committee in 2009, Watanabe is currently a visiting faculty member of Institute of Design Knowledge organised by Hong Kong Design Centre.

P. 208-209

TERADADESIGN

Established in 2003 by Naoki Terada, TERADADESIGN provides a wide range of design services including architectural, interior, furniture and product designs. Terada first graduated from Meiji University in Japan and later completed his studies at Architectural Association School of Architecture, London.

P. 025, 202-203

Torafu Architects

Founded in 2004 by Koichi Suzuno and Shinya Kamuro, Torafu Architects works on a diverse range of projects from architectural design, interior design, product design, spatial installations to filmmaking. The team has received numerous awards including the Design for Asia Grand Award in 2005 and the Grand Prize of the ELITA DESIGN AWARDS 2011.

P. 062-063

Whatshisname

Whatshisname is the work name of Sebastian Burdon, who was born in 1982 in Poland. First started out as an assistant to well-established British artists, Burdon creates to encourage the public to look at the surrounding world and to question it in a derisive, jeering way.

P. 146-147

Young & Innocent

An art and design partnership consisting of art director Lio Yeung and set designer Kay Au Yeung established in Hong Kong in 2012, Young & innocent is a creative platform reaching for aesthetic, highly-visualised, expressive and fashionable approach on design.

P. 012-013

Acknowledgements

We would like to thank all the designers and companies who have involved in the production of this book. This project would not have been accomplished without their significant contribution to the compilation of this book. We would also like to express our gratitude to all the producers for their invaluable opinions and assistance throughout this entire project. The successful completion also owes a great deal to many professionals in the creative industry who have given us precious insights and comments. And to the many others whose names are not credited but have made specific input in this book, we thank you for your continuous support the whole time.

Future Editions

If you wish to participate in viction:ary's future projects and publications, please send your website or portfolio to submit@victionary.com